Springer Theses

Recognizing Outstanding Ph.D. Research

Aims and Scope

The series "Springer Theses" brings together a selection of the very best Ph.D. theses from around the world and across the physical sciences. Nominated and endorsed by two recognized specialists, each published volume has been selected for its scientific excellence and the high impact of its contents for the pertinent field of research. For greater accessibility to non-specialists, the published versions include an extended introduction, as well as a foreword by the student's supervisor explaining the special relevance of the work for the field. As a whole, the series will provide a valuable resource both for newcomers to the research fields described, and for other scientists seeking detailed background information on special questions. Finally, it provides an accredited documentation of the valuable contributions made by today's younger generation of scientists.

Theses are accepted into the series by invited nomination only and must fulfill all of the following criteria

- They must be written in good English.
- The topic should fall within the confines of Chemistry, Physics, Earth Sciences, Engineering and related interdisciplinary fields such as Materials, Nanoscience, Chemical Engineering, Complex Systems and Biophysics.
- The work reported in the thesis must represent a significant scientific advance.
- If the thesis includes previously published material, permission to reproduce this must be gained from the respective copyright holder.
- They must have been examined and passed during the 12 months prior to nomination.
- Each thesis should include a foreword by the supervisor outlining the significance of its content.
- The theses should have a clearly defined structure including an introduction accessible to scientists not expert in that particular field.

More information about this series at http://www.springer.com/series/8790

Lei Zhang

Ultra-Broadly Tunable Light Sources Based on the Nonlinear Effects in Photonic Crystal Fibers

Doctoral Thesis accepted by
Tsinghua University, Beijing, China

Springer

Author
Dr. Lei Zhang
Tsinghua University
Beijing
China

Supervisor
Prof. Shizhong Xie
Tsinghua University
Beijing
China

ISSN 2190-5053   ISSN 2190-5061 (electronic)
Springer Theses
ISBN 978-3-662-48359-6   ISBN 978-3-662-48360-2 (eBook)
DOI 10.1007/978-3-662-48360-2

Library of Congress Control Number: 2015948733

Springer Heidelberg New York Dordrecht London

Printed on acid-free paper

Springer-Verlag GmbH Berlin Heidelberg is part of Springer Science+Business Media
(www.springer.com)

Parts of this thesis have been published in the following articles:

Zhang L, Yang S, Han Y, Chen H, Chen M, Xie S (2013) Optical parametric generation with two pairs of gain bands based on a photonic crystal fiber. Opt Commun 87: 22–26 (Reproduced with Permission).

Zhang L, Yang S, Han Y, Chen H, Chen M, Xie S (2013) Simultaneous generation of tunable giant dispersive waves in the visible and mid-infrared regions based on photonic crystal fibers. J Opt 15: 075201.

Zhang L, Yang S, Li P, Wang X, Gou D, Chen W, Luo W, Chen H, Chen M, Xie S (2013) An all-fiber continuously time-dispersion-tuned picosecond optical parametric oscillator at 1 μm region. Opt Express 21: 25167–25173 (Reproduced with Permission).

Zhang L, Yang S, Wang X, Gou D, Li X, Chen H, Chen M, Xie S (2013) Widely tunable all-fiber optical parametric oscillator based on a photonic crystal fiber pumped by a picosecond ytterbium-doped fiber laser. Opt Lett 38: 4534–4537 (Reproduced with Permission).

Zhang L, Yang S, Chen H, Chen M, Xie S (2013) Ultraviolet-shift supercontinuum generation by cross-phase modulation in photonic crystal fiber. CLEO: JW2A.13 (Reproduced with Permission).

Zhang L, Yang S, Chen H, Chen M, Xie S (2013) Wavelength-tunable red-shift Cherenkov radiation in photonic crystal fibers for mid-infrared wavelength generation. OECC: WS4-4.

Zhang L, Yang S, Wang X, Gou D, Chen H, Chen M, Xie S (2014) Demonstration of optical parametric gain generation at 1 μm regime based on a photonic crystal fiber pumped by picosecond mode-locked ytterbium-doped fiber laser. J Opt 16: 015202.

Zhang L, Yang S, Wang X, Gou D, Chen H, Chen M, Xie S (2014) Picosecond photonic crystal fiber-based doubly resonant optical parametric oscillator. IEEE Photon Technol Lett 26: 682–685.

Zhang L, Yang S, Wang X, Gou D, Chen H, Chen M, Xie S (2014) High-efficiency all-fiber optical parametric oscillator based on photonic crystal fibers pumped by ytterbium-doped fiber laser. Electron Lett 50: 624–626.

Zhang L, Yang S, Wang X, Gou D, Chen W, Luo W, Chen H, Chen M, Xie S (2014) Photonic crystal fiber based wavelength-tunable optical parametric amplifier and picosecond pulse generation. IEEE Photonics J 6: 1501908.

Zhang L, Yang S, Chen H, Chen M, Xie S (2014) Broadly time-dispersion-tuned narrow linewidth all-fiber-integrated optical parametric oscillator. OFC: W4E.2 (Reproduced with Permission).

Supervisor's Foreword

Novel wavelength fiber lasers are required urgently in a lot of areas such as biomedicine, remote sensing, and microscopy. However, the current fiber lasers are limited in several specific wavelength bands. In order to expand the operating wavelength range, new rare earth ion-doped materials are needed to be explored. However, this is difficult and expensive. The alternative method is to convert the fine developed laser power into the novel wavelength band via the nonlinear effect in optical fibers. Fiber optical parametric amplification (FOPA), also called four-wave mixing (FWM), has high gain and broad bandwidth. By utilizing the parametric gain provided by the FOPA, the fiber optical parametric oscillator (FOPO) can be built up. Specifically, the OPA is usually generated by pumping near the zero-dispersion wavelength (ZDW) of the optical fiber. However, the ZDW of the traditional fiber is limited in the telecommunication band. Thanks to the appearance of the photonic crystal fibers (PCFs), the ZDW can cover a wide wavelength range from the visible to the infrared band.

In the past four years, Lei Zhang has investigated intensively the generation of widely tunable light in nonconventional wavelength band based on PCFs. The highly-nonlinear PCFs with customized dispersion properties are designed and fabricated domestically. Widely tunable FOPA and FOPO are developed in the PCFs by pumping near 1.06 μm. Moreover, the energy conversion efficiency of the PCF-based OPO is largely improved by the customized cavity structures. The work in this thesis has improved the performances of the PCF-based OPO significantly. They are more reliable, powerful, and flexible than ever. In addition, the large-span FWM is achieved in a customized PCF with two ZDWs. Meanwhile, widely tunable dispersive waves can be generated in the visible and mid-infrared regions, and the ultraviolet (UV) light in deep UV band can be generated by the cross-phase modulation between the anti-Stokes signal and the Raman soliton. Overall, his work in this thesis has proved that PCF-based nonlinear generation is promising for the nonconventional wavelength band light source and it contains valuable information on the PCF-based OPA and OPO which I hope the readers will enjoy.

Beijing   Prof. Shizhong Xie
June 2015

Acknowledgements

Foremost, I offer my sincerest gratitude to my supervisor Prof. Shizhong Xie for the guidance and encouragement throughout my Ph.D. study and research, for his patience, enthusiasm, and profound knowledge. I learned a lot from Prof. Xie not only technical skills but also his rigorous scientific approach, which will benefit my career the rest of my life.

I would especially like to thank Dr. Sigang Yang for being my major advisor. I am deeply grateful to him for the discussions and analysis that helped me sort out the technical details of my work. I am also thankful to him for encouraging the use of correct grammar and consistent notation in my writings and for carefully reading and commenting on countless revisions of this manuscript.

I want to thank Prof. Yasutake Ohishi in the Research Center for Advanced Photon Technology of Toyota Technological Institute for understanding and support of my thesis translation. I work as a postdoctoral fellow in Prof. Ohishi's laboratory now. Without the comfortable and enjoyable atmosphere created by them, I would not complete the thesis translation as soon as possible.

I want to thank Dr. Wei Chen, and Dr. Wenyong Luo in FiberHome Telecommunication Technologies CO. LTD. for the fiber manufacturing during the research. The photonic crystal fibers with the zero-dispersion wavelength located around 1064 nm are provided by them.

I would also like to thank Dr. Jinyan Li in Huazhong University of Science and Technology for fiber manufacturing during the research. Several photonic crystal fiber samples with two zero-dispersion wavelengths are provided by him.

I want to thank Prof. Lantian Hou, Dr. Ying Han, Prof. Guiyao Zhou, and Prof. Shuguang Li at Yanshan University for providing the Ti: sapphire laser.

I want to thank Dr. Qing Sun and Dr. Yuqiang Deng in the National Metrology Institute of China for the measurement of the fiber dispersion.

I want to thank Prof. Minghua Chen, associate professor Hongwei Chen, associate professor Wei Zhang, and Dr. Qiang Zhou for their help and kind advices.

I would like to thank my research fellow students Xiaojian Wang, Doudou Gou, Fangjian Xing, Feifei Yin, Cheng Lei, Ruiyue Li, Ying Yu, Hongbiao Gao, for sharing their experiences and creating such an enjoyable environment in our laboratory.

Most importantly, I want to express my deepest gratitude to my parents, for their lifelong supporting and love. I would also like to thank to my beloved wife, Yue Xing. Thanks for the encouragement from her.

Finally, I appreciate the National Basic Research Program of China (973 Program) under Contract 2010CB327606, the National Nature Science Foundation of China under Contract . 61108007, and the Opened Fund of the State Key Laboratory on Integrated Optoelectronics for the financial support during my Ph.D. study.

Contents

Nomenclature

a	Coefficient
A	Mode area (m^2)
b	Coefficient
c	Coefficient
C	Conventional wavelength, 1530–1565 nm
	Compression ratio
D	Dispersion (ps/nm/km)
E	Electric field (N/C)
g	Gain factor
G	Gain (dB)
H	Bessel function of the third kind
J	Bessel function of the first kind
k	Wave number
L	Length (m)
	Long wavelength, 1565–1625 nm
n	Refractive index
P	Power (W)
R	Raman response
T,t	Time (s)
W	Pulse width (ps)
R	Repetition rate

Greek Symbols

α	Loss (dB/km)
β	Dispersion coefficient
	Mode-propagation constant
γ	Nonlinear coefficient ($W^{-1}\ km^{-1}$)
Δ	Difference
η	Duty cycle

κ Nonlinear phase mismatch (m^{-1})
λ Wavelength (nm)
μ Micro
v Velocity (m/s)
Φ Phase shift (°)
φ Phase (°)
ω Angular frequency (rad/s)
Ω Frequency detuning (Hz)

Scripts

a   Air
as   Amplified signal
c   Central
d   Dispersive wave
eff   Effective
g   Group
NL   Nonlinear
p   Photonic crystal fiber
        Pump
ras   Revised amplified signal
Re   Real part
s   Single mode fiber
        Signal
        Soliton
        Stokes
th   Threshold
z   Longitudinal component

Chapter 1
Introduction

1.1 Background

Lasers have attracted significant attentions since the first advent in 1960 due to their remarkable properties, such as monochromaticity, high coherence, good collimation, and high intensity. They have been widely used in a variety of fields, ranging from basic research to engineering applications, such as communications, industrial manufacture, and military affairs. Recently, they are also used in the emerging areas of marine science, atmospheric physics, nuclear fusion, and biological medicine [1, 2]. Their application fields are continuously expanded. The operating situations are becoming more and more diverse as well as complex. The laser technology must advance with times in order to satisfy the continuous emergence of the novel applications.

Conventionally, the commercially available lasers are classified into four major categories: gas lasers, solid-state lasers, dye lasers, and semiconductor lasers.

In the case of gas lasers, several kinds of gas can be selected as the gain medium. Benefiting from the high homogeneousness of the gas medium, high beam quality can be realized easily. However, due to the low density of gas medium, it is hard to achieve a significant population inversion in unit volume, which leads to the low gain and the limited output power.

In the case of solid-state lasers, optical crystals or optical glasses doped with impurity ions are used as the gain medium. When doped optical crystals are used, lasers with narrow linewidth can be obtained. In the case of doped optical glasses, since the gain material with large size is available, lasers with high energy and high peak power can be obtained easily. However, most of the optical crystals or glasses need to be operated in low-temperature environment. Complex water cooling systems are required, so that traditional solid-state lasers usually look bulky and cumbersome, and the operation is difficult [3].

© Springer-Verlag Berlin Heidelberg 2016
L. Zhang, *Ultra-Broadly Tunable Light Sources Based on the Nonlinear Effects in Photonic Crystal Fibers*, Springer Theses,
DOI 10.1007/978-3-662-48360-2_1

For dye lasers, organic dye-doped liquid solvents are used as the gain medium. The output wavelength can cover a broadband from ultraviolet to near-infrared. However, the practical applications are seriously limited by the instability of the liquid solvents.

Semiconductor lasers employ semiconductor material as the gain medium. They are widely used in a lot of areas due to the properties of high efficiency, small size, and easy to integrate.

1.2   Fiber Lasers

Recently, fiber lasers have received considerable attentions due to their ultra-fast and high-power characteristics. The typical setup is shown in Fig. 1.1. Generally, the intra-cavity gain is provided by a rare earth-doped fiber, and the pump can be coupled into the laser cavity by a wavelength division multiplex (WDM) fiber coupler. Usually, a ring cavity structure can be adopted in order to introduce optical feedback to the gain segment. The unidirectional transmission is guaranteed by inserting an optical isolator (ISO) into the cavity. The polarization state of output laser beam can be tuned by an intra-cavity polarization controller (PC). The oscillating light beam can be restricted in the fiber core for a long-distance trans-mission, so fiber lasers usually have high beam quality. With the fiber laser systems built by polarization maintaining fibers and polarization maintaining devices, the polarization state in the feedback cavity will be immune to the environmental variation, and the steady-state operation can be maintained in varieties of harsh conditions. In the high-power fiber laser systems, large-mode-area fibers can be used to decrease the effect of nonlinearity. In addition, the fiber output port can be connected with varieties of other fiber devices. In view of the advantages, several directions have been developed, such as high-power, multi-wavelength, narrow linewidth, few cycles, and ultrashort pulses [4, 5].

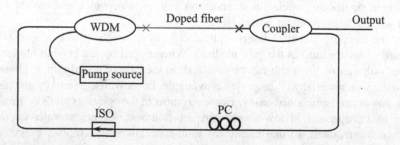

Fig. 1.1 The basic schematic of the fiber lasers

1.2.1 Background and Tendency of Fiber Lasers

The optical fibers have received a lot of attentions and a rapid development during the first several years. That is because the fused silica material has very low loss in the telecommunication band, and the silica optical fibers have the potential to transmit optical signals quickly [6]. Subsequently, the scientist found that the waveguide structure of optical fibers can restrain light successfully, so that perfect output light beam quality can be achieved. Since then, many scientists dedicated to improve the fiber performances. In 1970, the first single-mode silica fiber with low loss was fabricated by Corning. In 1985, the first low-loss silica fiber doped with Nd ions was drawn by the research group at the University of Southampton. The first fiber laser was built by using this Nd-doped fiber as the gain medium, and the single-mode output was realized at 1059 nm [7, 8]. In 1987, the Er-doped silica fiber was successfully fabricated by the same group. The first fiber amplifier was built by using the Er-doped fiber as the gain medium [9, 10]. Afterward, the fiber lasers and fiber amplifiers received more attentions and faster developments.

In 1988, Snitzer et al. [11] built a fiber laser by using the double-cladding Ed-doped fiber as the gain medium. Then, the high-power fiber lasers were pushed forward on the fast track. In 1990, the output power of fiber lasers was limited to several watts [12]. In 1999, the output power was increased to 100 W [13]. In 2002, the output power reached the one thousand level [14]. By 2009, the output power have reached 10,000 W [15]. The maximal output power was refreshed quickly. This is attributed to the mature fiber fabrication technology. In the case of pulsed fiber lasers, the maximal average output power was also refreshed frequently. In 2005, the average output power was in the level of 100 W at the maximum [16]. By 2009, the maximal average output power was enhanced to 1000 W [17]. The maximal output power increased exponentially.

Multi-wavelength fiber lasers have important applications in the fields of WDM communication and sensing. They can also be used to test the performance of the WDM devices. In 1994, Takahashi et al. reported a four-wavelength fiber laser settled with an arrayed waveguide grating (AWG) and four segments of Er-doped gain fibers, in which the interval between the adjacent wavelengths was 1.6 nm [18]. Later, Miyazaki et al. demonstrated a multi-wavelength fiber laser with 15 wavelengths emitted simultaneously. The wavelength interval was 1.6 nm. Two AWGs and 15 segments of Er-doped gain fiber were used in the system. Immediately after each segment of the Er-doped gain fiber, an optical ISO was used to prevent the unwanted reflection and an adjustable attenuator was applied to tune the power at each wavelength. Flat output power profile can be achieved [19]. In 1997, Hübner et al. proposed to realize multi-wavelength output by using only one segment of gain fiber. Experimentally, a distributed feedback (DFB) fiber laser was achieved by using a section of Er-doped fiber with a series of Bragg grating to

provide gain. Within the wavelength range of 4.2 nm, five wavelengths were
emitted simultaneously [20]. Bellemare et al. achieved multi-wavelength output by
using the acousto-optic modulation method to produce periodic frequency shift
[21]. In 2009, X.S. Liu et al. demonstrated a multi-wavelength fiber laser by
combining the nonlinear polarization rotation with the four-wave mixing
(FWM) effect. It can be operated with 38 wavelengths simultaneously. The interval
between each adjacent two wavelengths is 0.4 nm [22]. The multi-wavelength fiber
lasers will continue to be developed to obtain single-longitudinal-mode
(SLM) output at more wavelengths with similar intensities.

Narrow linewidth fiber lasers are urgently needed in the applications of coherent
communication, remote sensing, and fiber-optic interferometric sensor, in which
highly coherent light sources are needed. In the early years, the output 3-dB
bandwidth was about 10 nm [23]. Later, the scientist found that narrowband
reflection spectrum with 0.1-nm bandwidth can be obtained by using fiber Bragg
grating (FBG). The FBG and saturable absorber can be combined to build narrow
linewidth fiber laser [24]. Then, narrow linewidth was realized by using bandpass
filter formed by combination of multiple ring cavities [25]. However, mode hopping
induced by thermal noise was hard to avoid, because the fiber length included in the
laser cavity was at least several meters. In 2010, M.A. Quintela et al. wrote the
Bragg grating into a 50-cm-long Er-doped fiber. An exact SLM fiber laser was
achieved by avoiding the mode hopping with linear cavity structure formed by the
50-cm-long Er-doped fiber [26].

The primal motivation of developing ultra-fast fiber lasers is to serve as the
signal source in the telecommunication system. The first pulsed fiber laser was built
successfully as early as 1966 [27]. Later, the ultra-fast fiber lasers became more
arrestive because of the properties of high peak power and wide spectral profile.
Varieties of new technologies have been emerging, such as Q-switching and mode
locking.

1.2.2   New Wavelength Fiber Lasers

Presently, the fiber lasers mainly employ the doped fibers as the gain medium. They
can only be operated at several special wavelength regions, such as ~ 1060, ~ 1550,
and ~ 2000 nm. However, fiber lasers operated at new wavelength regions are
needed in many newly emerging applications. The working wavelength region of the
light source in many applications can determine the system performance, for
example, in the optical coherence tomography. How to extend the operating
wavelength region of the fiber lasers has been one of the most popular research fields
in recent years.

If doped fibers are continuously employed as the gain medium to develop fiber
lasers operated at new wavelength band, novel rare earth ions which can doped into

the fibers need to be explored. Many well-known groups from around the world have focused in this field for decades. It is a hard work to continuously find new rare earth ions emitted at new wavelength regions. An alternative method is to convert the well-developed lasers to new wavelength regime by exploring the nonlinear effects in optical fibers.

Raman fiber lasers can emit at some wavelength regions where conventional fiber laser cannot oscillate. One of the main features is the broad Raman gain band. If the pump power is strong enough, multi-cascaded Raman gain bands can be used. The gain band of doped fiber can generally cover dozens of nanometers. Cascaded Raman gain can cover hundreds of nanometers, and the gain region can be tuned by simply adjusting the pump wavelength. The Raman amplifiers have been reported by pumping at 1.3, 1.52, and 1.65 μm, respectively. Continuous wave and pulsed Raman fiber lasers can be realized by utilizing the gain provided by the Raman amplifiers [28, 29]. A Raman laser operated at 1120 nm has been demonstrated by using the Yb-doped fiber laser as the pump.

Although Raman fiber lasers have the potential to emit at any wavelength, the application is limited by the strict requirements of Raman amplifiers. A long length of optical fibers is needed, distributed pumps are required in order to reduce the effects of Brillouin scattering, and a pump source operated at the suitable wavelength region often is difficult to obtain.

Except for stimulated Raman scattering (SRS) effect, the four-wave mixing (FWM) in optical fiber can also provide gain at new wavelength band. FWM phenomenon can be usually observed for pump wavelength close to the zero-dispersion wavelength of gain fiber. When the pump wavelength is tuned within the normal dispersion wavelength regime, narrow gain sidebands with large wavelength shift from the pump can be obtained. In Ref. [30], the FWM idler sideband covers a wide band over 200 nm with the pump wavelength 20 nm shorter than the zero-dispersion wavelength.

Based on the FWM effect in optical fibers, a small signal located in the FWM sidebands can be amplified by receiving energy from the pump. This process was named as fiber optical parametric amplifier (FOPA). When the pump wavelength is selected close to the zero-dispersion wavelength of the gain fibers, high gain can be obtained in a wide wavelength range. Torounidis et al. [31] reported the OPA gain over 10 dB was achieved from 1510 to 1610 nm in a 100-m-length HNLF.

With a feedback optical path added to the parametric amplification process, small signals from the FWM sidebands can be transmitted and amplified. After multiple circulation, a dynamic balance can be built up, and thus, a stable output at new wavelength can be obtained, which was called fiber-optical parametric oscillator (FOPO) and can be used to build new wavelength fiber lasers. FWM-based fiber laser has been realized from 1400 to 1700 nm in a highly-nonlinear dispersion-shifted fiber by pumping near the zero-dispersion wavelength [32].

1.3   Nonlinear Effects in Fibers

The evolution of optical pulses in the medium of optical fiber with third-order nonlinear effects can be expressed by the nonlinear Schrödinger equation (NLSE) [33].

$$\frac{\partial E}{\partial z} = -\frac{\alpha}{2}E - \sum_{m=2}^{M} i^{m-1} \frac{\beta_m}{m!} \frac{\partial^m E}{\partial t^m} + i\left(\gamma + i\gamma_1 \frac{\partial}{\partial t}\right)\left[E(z,t)\int R(t')|E(z,t-t')|dt'\right] \quad (1.1)$$

where E is the envelope of the complex field, β_m is the m-order dispersion coefficient at the carrier frequency, γ represents the nonlinear coefficient, and γ_1 is the derivative of the nonlinear coefficient with respect to the frequency. $R(t) = f_a h_a(t)$, where $f_a = 0.75$ represents the contribution of the Raman response to the nonlinear process, $h_a(t)$ is the Raman response function for the silica material, which can be expressed as follows[34]:

$$h_a(t) = \frac{t_1^2 + t_2^2}{t_1 t_2^2}\exp(-t/t_2)\sin(t/t_1) \quad (1.2)$$

where t_1 and t_2 are usually set to be 12.2 and 32 fs. The fiber attenuation, group velocity dispersion, self-phase modulation, and SRS are considered in the NLSE, and the pulse evolution can be estimated accurately.

1.3.1   Self-Phase Modulation

The relationship between the refractive index and the transmitted light intensity can be expressed as follows:

$$\tilde{n}(\omega, |E|^2) = n(\omega) + n_2|E|^2 \quad (1.3)$$

where $n(\omega)$ is the linear term, which is independent with the light intensity and n_2 is the nonlinear refractive index of the fiber material. It is associated with the real part of the third-order susceptibility [33].

$$n_2 = \frac{3}{8n}\text{Re}(\chi^{(3)}) \quad (1.4)$$

The value of n_2 is usually very small. However, when the light intensity $|E|^2$ is large enough, the effect of n_2 cannot be ignored. When the light signal is transmitted in optical fibers, self-induced phase shift can be generated, which is named as self-phase modulation, and can be expressed as follows [33]:

$$\phi = \tilde{n}k_0 L = (n + n_2|E|^2)k_0 L \quad (1.5)$$

where L is fiber length and $k_0 = 2\pi/\lambda$. The term related to the light intensity is defined as nonlinear phase shift.

$$\phi_{NL} = n_2|E|^2 k_0 L \tag{1.6}$$

Frequency chirp can be introduced to the original pulse by the time-dependent nonlinear phase shift [33]:

$$\delta\omega(t) = -\partial\phi_{NL}/\partial t \tag{1.7}$$

The new frequency component introduced by self-phase modulation usually locates near the original frequency.

1.3.2 Cross-Phase Modulation

Cross-phase modulation refers to the phenomenon that phase shift is not only induced from the signal itself but also from other signals transmitted in the fiber simultaneously. When two optical signals at frequencies of ω_1 and ω_2 with the same polarization direction transmit simultaneously in a fiber, the total electrical field can be expressed as follows[33]:

$$E = \frac{1}{2}[E_1 \exp(-i\omega_1 t) + E_2 \exp(-i\omega_2 t) + c.c.] \tag{1.8}$$

where c.c. denotes the complex conjugate. The nonlinear phase shift sensed by the optical signal at ω_1 can be expressed as follows:

$$\phi_{NL} = n_2 k_0 L(|E_1|^2 + 2|E_2|^2) \tag{1.9}$$

where $2n_2 k_0 L|E_2|^2$ is from the cross-phase modulation between the signals at ω_1 and ω_2. It can be seen that the contribution of XPM to the phase shift is the twice times to that of the SPM. The XPM requires two signals overlap in time and space simultaneously. When two pulse trains are transmitted in fibers simultaneously, the walk-off effect should be considered [33].

1.3.3 Stimulated Raman Scattering

When high-power laser is transmit in silica optical fibers, a down frequency gain spectrum with a bandwidth of about 40 THz can be generated, and the frequency gap between the gain peak and the pump is about 13 THz. This is from the interaction between the pump photon and the silica material. From the point of view of quantum optics, the process can be described as follows: When the fused silica molecule receives a photon, it can jump from the ground state to the high-lying virtual state, then jump to the vibrational state by emitting a low-frequency photon, and then jump to the ground state by emitting an optical phonon. This process is called SRS. The 40-THz gain band is determined by the vibrational state with wide energy band [33].

The evolutions of the pump and the Stokes sideband generated by the SRS can be expressed by the coupled equations [33]:

$$\frac{dI_p}{dz} = -\frac{\omega_p}{\omega_s} g_R I_p I_s - \alpha_p I_p \tag{1.10}$$

$$\frac{dI_s}{dz} = g_R I_p I_s - \alpha_s I_s \tag{1.11}$$

where ω_p and ω_s are the frequencies of the pump and the Stokes sideband, respectively. I denotes the optical intensity, α is the loss coefficient, and g_R is the Raman gain coefficient. When $\alpha_p = \alpha_s = 0$, the Eqs. (1.10) and (1.11) can be combined to:

$$\frac{d}{dz}\left(\frac{I_s}{\omega_s} + \frac{I_p}{\omega_p}\right) = 0 \tag{1.12}$$

It is clear that the photon numbers are conserved when the optical loss is ignored in the SRS process. Every vanished pump photon will generate a Stokes photon.

To analyze the optical intensity of the Stokes sideband substitutes the solution of Eq. (1.10) into Eq. (1.11)

$$\frac{dI_s}{dz} = g_R I_0 \exp(-\alpha_p z) I_s - \alpha_s I_s \tag{1.13}$$

where I_0 denotes the pump intensity at $z = 0$. The solution of Eq. (1.13) can be expressed as follows:

$$I_s(L) = I_s(0) \exp(g_R I_0 L_{eff} - \alpha_s L) \tag{1.14}$$

where $I_s(0)$ is the optical intensity of the Stokes sideband at $z = 0$, L is the fiber length, and L_{eff} is the effective fiber length, which is given by:

$$L_{eff} = [1 - \exp(-\alpha_p L)]/\alpha_p \tag{1.15}$$

The power of the Stokes sideband can be calculated by an integral for the frequency:

$$P_s(L) = \int_{-\infty}^{\infty} \hbar\omega \exp[g_R(\omega_p - \omega)I_0 L_{eff} - \alpha_s L]d\omega \tag{1.16}$$

After substituting the dependence between g_R and frequency into Eq. (1.16), utilizing the steepest descent method, the power of the Stokes wave can be calculated to be as follows:

$$P_s(L) = P_{s0}^{eff} \exp[g_R(\Omega_R)I_0 L_{eff} - \alpha_s L] \qquad (1.17)$$

where the initially effective incident signal power is given as follows:

$$P_{s0}^{eff} = \hbar\omega_s B_{eff} \qquad (1.18)$$

$$B_{eff} = \left(\frac{2\pi}{I_0 L_{eff}}\right)^{1/2} \left|\frac{\partial^2 g_R}{\partial\omega^2}\right|_{\omega=\omega_s}^{-1/2} \qquad (1.19)$$

B_{eff} is the bandwidth near the wavelength with a maximal gain. When the residual pump power is similar with the Stokes power at the output, this incident pump power is defined as the Raman threshold power, which can be estimated by the equation:

$$P_{th} \approx \frac{16A_{eff}}{g_R L_{eff}} \qquad (1.20)$$

where A_{eff} is the effective mode field area of the fundamental mode in optical fiber:

$$A_{eff} = \frac{\left(\int\int\limits_{-\infty}^{\infty} |F(x,y)|^2 dxdy\right)^2}{\int\int\limits_{-\infty}^{\infty} |F(x,y)|^4 dxdy} \qquad (1.21)$$

where $F(x, y)$ is the mode field distribution. When the pump power exceeds the Raman threshold power, the optical intensity of the Stokes sideband will increase exponentially [33].

1.3.4 Four-Wave Mixing

FWM refers to an energy exchange phenomenon between four photons with the momentum and energy conserved. In optical fibers, the FWM process will modulate the fiber parameters, so it is also referred to as parametric process. During this process, two pump photons are annihilated, a signal photon and an idler photon will be generated simultaneously. An exception is that three photons exchange the energies to one photon, which is named as third harmonic generation (THG). In optical fibers, the phase-matching condition of THG is hard to satisfy, and the efficiency is very low [33]. In this thesis, we mainly focus on the common FWM, where the phase-matching condition is relatively easy to satisfy.

To meet the law of conservation of energy, the frequencies of the four photons should satisfy the equation of:

$$\omega_1 + \omega_2 = \omega_3 + \omega_4 \qquad (1.22)$$

To meet the law of conservation of momentum, the propagation constants of the four photons should satisfy the phase-matching condition of:

$$\Delta k = \beta_3 + \beta_4 - \beta_1 - \beta_2 = (n_3\omega_3 + n_4\omega_4 - n_1\omega_1 - n_2\omega_2)/c = 0 \qquad (1.23)$$

where c denotes the velocity of light and n denotes the effective index. After the phase-matching condition satisfied, the signal and idler waves can be generated from optical noise by the large power pump, and this process is named as optical parametric generation. When a small signal is coupled into the fiber with the pump, it can be amplified significantly, and the idler can also be generated, and this process is called as optical parametric amplification.

During the FWM process, the evolution of the optical intensities can be described by the equations of [33]:

$$\frac{dA_1}{dz} = \frac{in_2\omega_1}{c}\left[\left(f_{11}|A_1|^2 + 2\sum_{k\neq 1}f_{1k}|A_k|^2\right)A_1 + 2f_{1234}A_2^*A_3A_4e^{i\Delta kz}\right] \qquad (1.24)$$

$$\frac{dA_2}{dz} = \frac{in_2\omega_2}{c}\left[\left(f_{22}|A_2|^2 + 2\sum_{k\neq 2}f_{2k}|A_k|^2\right)A_2 + 2f_{2134}A_1^*A_3A_4e^{i\Delta kz}\right] \qquad (1.25)$$

$$\frac{dA_3}{dz} = \frac{in_2\omega_3}{c}\left[\left(f_{33}|A_3|^2 + 2\sum_{k\neq 3}f_{3k}|A_k|^2\right)A_3 + 2f_{3412}A_1A_2A_4^*e^{-i\Delta kz}\right] \qquad (1.26)$$

$$\frac{dA_4}{dz} = \frac{in_2\omega_4}{c}\left[\left(f_{44}|A_4|^2 + 2\sum_{k\neq 4}f_{4k}|A_k|^2\right)A_4 + 2f_{4312}A_1A_2A_3^*e^{-i\Delta kz}\right] \qquad (1.27)$$

where the overlap integral can be expressed as follows:

$$f_{jk} = \frac{\left\langle |F_j|^2|F_k|^2\right\rangle}{\left\langle |F_j|^2\right\rangle\left\langle |F_k|^2\right\rangle} \qquad (1.28)$$

$$f_{ijkl} = \frac{\left\langle F_i^*F_j^*F_kF_l\right\rangle}{\left[\left\langle |F_i|^2\right\rangle\left\langle |F_j|^2\right\rangle\left\langle |F_k|^2\right\rangle\left\langle |F_l|^2\right\rangle\right]^{1/2}} \qquad (1.29)$$

where the angle brackets denote to integrate over the Cartesian coordinates x and y. It is very difficult to obtain the accurate solutions by the analytic methods. To analyze the evolution of the optical intensities at the four different frequencies in analytics, some simplification and hypothesis should be taken. In the FWM process, the pump power usually much larger than the signal and idler power, so the loss of the pump wave can be neglected. The overlap integral terms can be supposed to be approximately equal.

$$f_{ijkl} \approx f_{ij} \approx 1/A_{\text{eff}} \quad (i,j,k,l = 1,2,3,4) \tag{1.30}$$

After ignoring the differences between the four frequencies, the nonlinear coefficients can be expressed as follows:

$$\gamma_j = n_2 \omega_j / (c A_{\text{eff}}) \approx \gamma \tag{1.31}$$

The solutions of Eqs. (1.24) and (1.25) can be expressed as follows:

$$A_1(z) = \sqrt{P_1} \exp[i\gamma(P_1 + 2P_2)z] \tag{1.32}$$

$$A_2(z) = \sqrt{P_2} \exp[i\gamma(P_2 + 2P_1)z] \tag{1.33}$$

It can be seen that the pump wave only affected by the SPM and XPM effects. Substitute Eqs. (1.32) and (1.33) into Eqs. (1.26) and (1.27):

$$\frac{dA_3}{dz} = 2i\gamma \left[(P_1 + P_2)A_3 + \sqrt{P_1 P_2} e^{-i\theta} A_4^* \right] \tag{1.34}$$

$$\frac{dA_4^*}{dz} = -2i\gamma \left[(P_1 + P_2)A_4^* + \sqrt{P_1 P_2} e^{i\theta} A_3 \right] \tag{1.35}$$

where $\theta = [\Delta k - 3\gamma(P_1 + P_2)]z$. Then, introduce the equation as follows:

$$B_j = A_j \exp[-2i\gamma(P_1 + P_2)z] \quad (j = 3,4) \tag{1.36}$$

According to Eqs. (1.34)–(1.36), we can obtain the following expressions:

$$\frac{dB_3}{dz} = 2i\gamma \sqrt{P_1 P_2} \exp(-i\kappa z) B_4^* \tag{1.37}$$

$$\frac{dB_4^*}{dz} = -2i\gamma \sqrt{P_1 P_2} \exp(i\kappa z) B_3 \tag{1.38}$$

where the effective phase mismatch can be expressed as follows:

$$\kappa = \Delta k + \gamma(P_1 + P_2) \tag{1.39}$$

Take differential to the left and right hand of the Eq. (1.37), the B_4^* can be canceled by Eq. (1.38), and an equation about B_3 can be obtained.

$$\frac{d^2 B_3}{dz^2} + i\kappa \frac{dB_3}{dz} - (4\gamma^2 P_1 P_2) B_3 = 0 \tag{1.40}$$

Using the same method, an equation about B_4^* can be obtained. The solution can be expressed as follows:

$$B_3(z) = (a_3 e^{gz} + b_3 e^{-gz}) \exp(-i\kappa z/2) \tag{1.41}$$

$$B_4^*(z) = (a_4 e^{gz} + b_4 e^{-gz}) \exp(i\kappa z/2) \tag{1.42}$$

where the coefficients of a_3, b_3, a_4, and b_4 are determined by the boundary condition, and the parametric gain is determined by the pump power.

$$g = \sqrt{(\gamma P_0 r)^2 - (\kappa/2)^2} \tag{1.43}$$

where r and P_0 can be expressed as follows:

$$r = 2(P_1 P_2)^{1/2}/P_0 \quad P_0 = P_1 + P_2 \tag{1.44}$$

It can be seen that the parametric gain has the maximal value when the effective phase mismatching is vanished. The FWM can be observed clearly when the phase-matching condition is satisfied. In optical fibers, the phase-matching condition can be expressed as follows:

$$\kappa = \Delta k_M + \Delta k_W + \Delta k_{NL} = 0 \tag{1.45}$$

where $\Delta k_M, \Delta k_W$ and Δk_{NL} represent the mismatching term induced by the material dispersion, waveguide dispersion, and the nonlinear effects, respectively. The adjustable dispersion and nonlinear properties can facilitate the phase matching realized in a large wavelength region.

1.4 Photonic Crystal Fibers

Photonic crystal fibers (PCFs), which are also known as microstructured optical fibers (MOFs) or holey fibers (HFs), are different from the conventional single-mode fibers fabricated by the chemical vapor deposition method. They are fabricated by the stack-and-draw or hot extrusion techniques. The cladding has a periodic refractive index distribution unchanged along the length of the fiber.

The fiber core is introduced by a defect destroyed the periodic cladding structure. The shape, size, and material composition of the fiber cladding and core can be varied. The special structure ensures the PCFs have a lot of special properties, such as endless single mode, adjustable dispersion, large mode area, high nonlinearity, and birefringence [35]. According to the light-guiding mechanism, the PCFs can be divided into two categories: index-guiding PCF and photonic bandgap PCF. In 1996, the index-guiding PCF was firstly fabricated by J.C. Knight et al. [36] at the University of Southampton. Then, the first photonic bandgap PCF was fabricated by R.F. Cregan et al. at the University of Bath in 1999. In the bandgap PCF, the light can be guided in the air core with low refractive index [37]. Subsequently, a lot of famous groups from different countries focus on to investigate this kind of new fiber. The fiber performances have been significantly improved, a lot of related technologies have been developed, and the application fields have been expanded continuously.

Not long after the invention of the silica PCF, several kinds of post-processing techniques were developed. A taped PCF can be formed by drawing an established PCF. At the taped waist, the core diameter can be narrowed to hundreds of nanometers [38]. The selected air holes in the fiber cladding or core can be filled with gas or liquid, the fiber dispersion can be tailored, and the transparent wavelength region can be enlarged [39]. The transparent wavelength band can be expanded to mid-infrared and THz region in the PCFs fabricated by nonsilica glass or polymer [40, 41]. All solid PCFs can also been fabricated by selecting different glass to form the periodic structure along the length of the fiber [42].

Due to the diversity and excellent performances of the PCFs, they are used as a key optical component in varieties of systems. Highly-nonlinear PCFs have been used in commercial supercontinuum (SC) light source [43]. Special PCFs with designed structure have been used as dispersion compensate devices in the telecommunication systems [44, 45]. Large-mode-area PCFs doped with rear earth ions have been used to build high-power fiber lasers [46, 47]. The PCFs have also been widely used in optical sensing and polarization maintaining systems [48, 49]. It is prospective that PCF can be used in more and more functional systems.

1.5 Motivation of Thesis

The application field of lasers has been expanded continuously, and the application environment becomes more and more diversity and complex. This requests a higher performance of the stability, power, and accessing wavelength regions. Fiber lasers are more attractive for the high output power and the stability in any kind of severe environments. However, the fiber lasers can only be operated in several wavelength bands. It is urgent and desired to extend the operating wavelength region of the fiber lasers. Basically, it is quite difficult to develop a new material that can emit in the nonconventional wavelength band. We selected to convert the finely developed laser power into a broadband novel wavelength region by utilizing the highly nonlinear property of the PCFs.

1.6   Thesis Organization

To achieve a widely tunable fiber laser with high energy conversion efficiency, PCFs with high nonlinearity and flexible dispersion are used as the nonlinear medium. We theoretically design the fiber structures and then send them to our partner group (FiberHome Telecommunication Technologies Co. Ltd.) for drawing the PCFs. Using these PCFs, we can investigate the nonlinear effects for emitting at new wavelength. The research content includes theoretical predicts, optical parametric testing, PCF-based optical parametric amplifier, and PCF-based optical parametric oscillator.

The organization of this thesis is listed as follows:

In Chap. 2, the dependences of the spectral properties of parametric gain on the dispersion, nonlinear, and pump wavelength are investigated theoretically. Subsequently, a PCF with two zero-dispersion wavelengths is designed precisely to exchange the signal at the telecommunication band to blue and green visible bands. The optical parametric properties are tested by using a Ti: sapphire laser as the pump. The dispersive wave and cross-phase modulation effects in the PCF are also investigated theoretically and experimentally.

In Chap. 3, a mode-locked ytterbium-doped fiber laser is used as the pump source. The PCFs with the zero-dispersion wavelengths located near 1060 nm are designed and fabricated. PCF-based optical parametric amplifier is constructed. An optimized theory to evaluate the gain of pulse-pumped FOPA is proposed. The properties of the pulse-pumped PCF-based optical parametric amplifier are investigated experimentally.

In Chap. 4, the parametric spectral components generated by the FWM in the PCF are fed back into the PCF through a ring cavity. The PCF-based optical parametric oscillators are constructed. The wavelength tuning mechanisms are analyzed. By tuning the pump wavelength, the PCF-based OPO can emit at a wide wavelength range. Using the time-dispersion-tuned technique, different wavelength components can be tuned to synchronize with the pump pulse by tuning the cavity dispersion. The output wavelength can be continuously tuned without tuning the pump wavelength. By combining the pump wavelength tuning method with the time-dispersion-tuned technique, the output wavelength can be continuously tuned in a wide wavelength range.

In Chap. 5, the factors to affect the energy conversion efficiency of the PCF-based optical parametric oscillator are analyzed. Special cavity structures are designed to improve the energy conversion efficiency. The signal synchronized with the pump pulse has larger intensity and narrower bandwidth than that of the new generated idler at the output port, so a double cavity structure is designed.

The energy conversion efficiency can be increased when the signal and idler synchronize with the pump simultaneously. For further increasing the conversion efficiency, an all fiber cavity is adopted to the PCF-based optical parametric oscillator.

In Chap. 6, the content and the innovation points are summarized.

References

1. J. Wilson, J.F.B. Hawkes, *Lasers: Principles and Applications* (Prentice Hall, London, 1987)
2. http://www.baike.com/wiki/%E6%BF%80%E5%85%89?prd=so_1_doc
3. R.C. Powell, *Physics of Solid-State Laser Materials* (Springer, Berlin, 1998)
4. C. Jauregui, J. Limpert, A. Tünnermann, High-power fiber lasers. Nat. Photon. **7**(11), 861–867 (2013)
5. M.E. Fermann, I. Hartl, Ultrafast fiber lasers. Nat. Photon. **7**(11), 868–874 (2013)
6. C.K. Kao, T.W. Davies, Spectroscopic studies of ultra-low loss optical glasses I: single beam method. J. Sci. Instrum. **2**(1), 1063–1068 (1968)
7. S.B. Poole, D.N. Payne, M.E. Fermann, Fabrication of low-loss optical fibers containing rare-earth inons. Electron. Lett. **21**(17), 737–738 (1985)
8. R.J. Mears, L. Reekie, S.B. Poole, D.N. Payne, Neodymium-doped silica single-mode fiber laser. Electron. Lett. **21**(17), 738–740 (1985)
9. R. Mears, L. Reekie, I.M. Jauncey, D.N. Payne, Low-noise erbium-doped fiber amplifier operating at 1.54 μm. Electron. Lett. **23**(19), 1026–1028 (1987)
10. E. Desurvire, J.R. Simpson, P.C. Becker, High-gain erbium-doped travelingwave fiber amplifier. Opt. Lett. **12**(11), 888–890 (1987)
11. E. Snitzer, H. Po, F. Hakimi, R. Tumminelli, B.C. McCollum. Double clad, offset core Nd fiber laser, in *Optical Fiber Sensors* (Optical Society of America, Washington 1988): PD5
12. J.D. Minelly, E.R. K. Taylor, P. Jedrzejewski, J. Wang, D.N. Payne. Laser-diode-pumped Nd-doped fibre laser with output power > 1 W, in *Laser and Electro-Optics*, 1992
13. V. Dominic, S. MacCormack, R. Waarts, S. Sanders, S. Bickness, R. Dohle, E. Wolak, P.S. Yeh, E. Zucker, 110 W fibre laser. Electron. Lett. **35**(14), 1158–1160 (1999)
14. Y. Jeong, J.K. Sahu, D.N. Payne, J. Nilsson, Ytterbium-doped large-core fiber laser with 1.36 kW of continuous-wave output power. Opt. Express **12**(25), 6088–6092 (2004)
15. V. Gapontsev, V. Fomin, A. Ferin, M. Abramov. Diffraction limited ultra-high-power fiber lasers, in *Advanced Solid-State Photonics* (Optical Society of America, Washington, 2010), AWA1
16. S.W.F. Röser, J. Rothhard, B. Ortac, A. Liem, O. Schidt, T. Schreiber, J. Limpert, A. Tünnermann, 131 W 220 fs fiber laser system. Opt. Lett. **30**(20), 2754–2756 (2005)
17. T. Eidam, S. Hanf, E. Seise, T.V. Andersen, T. Gabler, C. Wirth, T. Schreiber, J. Limpert, A. Tünnermann, Femtosecond fiber CPA system emitting 830 W average output power. Opt. Lett. **35**(2), 94–96 (2010)
18. H. Takahashi, H. Toba, Y. Inoue, Multiwavelength ring laser composed of EDFAs and an arrayed waveguide wavelength multiplexer. Electron. Lett. **30**(1), 44–45 (1994)
19. T. Miyazaki, N. Edagawa, S. Yamamoto, S. Akiba, A multiwavelength fiber ring-laser employing a pair of silica-based arrayed-waveguide-gratings. IEEE Photon. Technol. Lett. **9** (7), 910–912 (1997)
20. J. Hübner, P. Varming, M. Kristensen, Five wavelength DFB fibre laser source for WDM systems. Electron. Lett. **33**(2), 139–140 (1997)
21. A. Bellemare, M. Rochette, M. Têtu, S. LaRochelle. Multifrequency erbium-doped fiber ring lasers anchored on the ITU frequency grid, in *Optical Fiber Communication Conference, OFC*: 1999, TuB5

22. X.S. Liu, L. Zhan, X. Hu, H.G. Li, Q.S. Shen, Y.X. Xia, Multiwavelength erbium-doped fiber laser based on nonlinear polarization rotation assisted by four-wave-mixing. Opt. Commun. **282**(14), 2913–2916 (2009)

23. K. Liu, M. Digonnet, K. Fesler, B.Y. Kim, H.J. Shaw, Broadband diode-pumped fibre laser. Electron. Lett. **24**(14), 838–840 (1988)

24. Y.W. Song, S.A. Havstad, D. Starodubov, Y. Xie, A.E. Willner, J. Feinber, 40-nm-wide tunable fiber ring laser with single-mode operation using a highly stretchable FBG. IEEE Photon. Technol. Lett. **13**(11), 1167–1169 (2001)

25. C.H. Yeh, T.T. Huang, H.C. Chien, C.H. Ko, S. Chi, Tunable S-band erbium-doped triple-ring laser with single-longitudinal-mode operation. Opt. Express **15**(2), 382–386 (2007)

26. M.A. Quintela, R.A. Perez-Herrera, I. Canales, M. Fernandez-Vallejo, M. Lopez-Amo, J.M. López-Higuera, Stabilization of dual-wavelength erbium-doped fiber ring lasers by single-mode operation. IEEE Photon. Technol. Lett. **22**(6), 368–370 (2010)

27. J. DeMaria, D.A. Stetser, H. Heynau, Self mode-locking of lasers with saturable absorbers. Appl. Phys. Lett. **8**(7), 174–176 (1966)

28. Y. Feng, L.R. Taylor, D.B. Calia, 150 W highly-efficient Raman fiber laser. Opt. Express **17** (26), 23678–23683 (2009)

29. J. Schröder, S. Coen, F. Vanholsbeeck, T. Sylvestre, Passively mode-locked Raman fiber laser with 100 GHz repetition rate. Opt. let. **31**(23), 3489–3491 (2006)

30. M.E. Marhic, K.K.Y. Wong, L.G. Kazovsky, Wide-band tuning of the gain spectra of one-pump fiber optical parametric amplifiers. IEEE J. Sel. Top. Quant. Electron. **10**(5), 1133–1141 (2004)

31. T. Torounidis, P. Andrekson, Broadband single-pumped fiber-optic parametric amplifiers. IEEE Photon. Technol. Lett. **19**(9), 650–652 (2007)

32. Y. Zhou, K.K.Y. Cheung, S. Yang, P.C. Chui, K.K.Y. Wong, Widely tunable picosecond optical parametric oscillator using highly nonlinear fiber. Opt. Lett. **34**(7), 989–991 (2009)

33. G.P. Agrawal, *Nonlinear Fiber Optics* (Springer, Berlin Heidelberg, 2000)

34. Q. Lin, G.P. Agrawal, Raman response function for silica fibers. Opt. Let. **31**(21), 3086–3088 (2006)

35. P.S.J. Russell, Photonic-crystal fibers. J. Lightwave Technol. **24**(12), 4729–4749 (2006)

36. J.C. Knight, T.A. Birks, P.J. Russell, D.M. Atkin, All-silica single-mode optical fiber with photonic crystal cladding. Opt. Lett. **21**(19), 1547–1549 (1996)

37. R.F. Cregan, B.J. Mangan, J.C. Knight, T.A. Birks, P.J. Russell, P.J. Roberts, D.C. Allan, Single-mode photonic band gap guidance of light in air. Science **285**(5433), 1537–1539 (1999)

38. E. Mägi, P. Steinvurzel, B. Eggleton, Tapered photonic crystal fibers. Opt. Express **12**(5), 776–784 (2004)

39. F. Benabid, J.C. Knight, G. Antonopoulos, P.J. Russell, Stimulated Raman scattering in hydrogen filled hollow-core photonic crystal fiber. Science **298**(5592), 399–402 (2002)

40. V.V. Kumar, A. George, W. Reeves, J.C. Knight, P.J. Russell, F. Omenetto, A. Taylor, Extruded soft glass photonic crystal fiber for ultrabroad supercontinuum generation. Opt. Express **10**(25), 1520–1525 (2002)

41. H. Han, H. Park, M. Cho, J. Kim, Terahertz pulse propagation in a plastic photonic crystal fiber. Appl. Phys. Lett. **80**(15), 2634–2636 (2002)

42. F. Luan, A.K. George, T.D. Hedley, G.J. Pearce, D.M. Bird, J.C. Knight, P.J. Russell, All-solid photonic bandgap fiber. Opt. Lett. **29**(20), 2369–2371 (2004)

43. J.M. Dudley, G. Genty, S. Coen, Supercontinuum generation in photonic crystal fiber. Rev. Mod. Phys. **78**(4), 1135 (2006)

44. S. Yang, Y. Zhang, L. He, S. Xie, Broadband dispersion-compensating photonic crystal fiber. Opt. let. **31**(19), 2830–2832 (2006)

45. L. Han, L. Liu, Z. Yu, H. Zhao, X. Song, J. Mu, X. Wu, J. Long, X. Liu, Dispersion compensation properties of dual-concentric core photonic crystal fibers. Chin. Opt. Lett. **12**(1), 010603 (2014)

46. X. Fang, M. Hu, C. Xie, Y. Song, L. Chai, C. Wang, High pulse energy mode-locked multicore photonic crystal fiber laser. Opt. Let. **36**(6), 1005–1007 (2011)
47. K. Guo, X. Wang, C. Luo, P. Zhou, B. Shu, Analysis of the maximum extractable power of photonic crystal fiber lasers. Chin. Opt. Lett. **12**(s2), 21411 (2014)
48. H. Liang, W. Zhang, P. Geng, Y. Liu, Z. Wang, J. Guo, S. Gao, S. Yan, Simultaneous measurement of temperature and force with high sensitivities based on filling different index liquids into photonic crystal fiber [J]. Opt. Lett. **38**(7), 1071–1073 (2013)
49. Z. Wang, G. Ren, S. Lou, S. Jian, Supercell lattice method for photonic crystal fibers [J]. Opt. Express **11**(9), 980–991 (2003)

Chapter 2
New Wavelength Generation Based on PCF with Two Zero-Dispersion Wavelengths (TZDWs)

2.1 Multipole Method

Since the PCF has flexible cladding structures, the simulation method for conventional optical fibers cannot be used for the evaluation of the PCF characteristics accurately. In recent years, several methods have been developed to improve the numerical precision, such as the plane wave expansion method, beam propagation method, finite element method, and multipole method. In this thesis, we mainly use the multipole method to estimate the fiber performances.

Compared with other methods, the limited cladding structure is considered in the multipole algorithm, which is especially suitable for the microstructured fibers with circle air holes distributed in the cladding. The calculation about the mode-propagation constant for the PCFs with complex structure can be finished more accurately and faster than other methods.

In the multipole method, every air hole in the cladding is considered as a scattering cell. The electromagnetic field components can be expressed by the Bessel functions in the cylindrical coordinate system. The solution of the Helmholtz equation can be obtained by using the boundary conditions. The longitudinal component of the electric field in the nth air hole can be expressed as follows [1, 2]:

$$E_z = \sum_{m=-\infty}^{\infty} a_m^{(n)} J_m\left(k_\perp^i r_n\right) \exp(im\varphi_n) \exp(i\beta z) \tag{2.1}$$

The longitudinal component of the electric field in the substrate material around the nth air hole can be expressed as follows:

$$E_z = \sum_{m=-\infty}^{\infty} \left[b_m^{(n)} J_m\left(k_\perp^e r_n\right) + c_m^{(n)} H_m^n\left(k_\perp^e r_n\right) \right] \times \exp(im\phi_n) \exp(i\beta z) \tag{2.2}$$

© Springer-Verlag Berlin Heidelberg 2016
L. Zhang, *Ultra-Broadly Tunable Light Sources Based on the Nonlinear Effects in Photonic Crystal Fibers*, Springer Theses,
DOI 10.1007/978-3-662-48360-2_2

where

$$k_\perp^i = (k_0^2 n_i^2 - \beta^2)^{1/2} \quad k_\perp^e = (k_0^2 n_e^2 - \beta^2)^{1/2} \tag{2.3}$$

where $n_i = 1$ denotes the refractive index of the air, n_e denotes the refractive index of the fused silica, $k_0 = 2\pi/\lambda$ is used to express the wave numbers in the free space, and β denotes the mode-propagation constant. The magnetic field components have the similar expression with the electric field components. The coefficients of a_m, b_m, and c_m can be obtained by using the boundary condition of the electromagnetic field. The effective refractive index n_{eff} and the effective mode area A_{eff} can be obtained by using the expression between the mode-propagation constant and the wave numbers in the free space $\beta = n_{\text{eff}} k_0$. Then, the nonlinear coefficient, group velocity dispersion, loss, and birefringence can be obtained.

2.2 Parametric Amplification Based on PCF with TZDW

2.2.1 The Design and Fabrication of the PCF with TZDW

The ZDW is an important parameter for optical fibers. When the pump is located near the ZDW, the nonlinear effects can be increased remarkably, and the optical parametric components can be generated easily. Fiber-based optical parametric generation is the cornerstone of the FOPA, FOPO, and all-optical wavelength converter. The ZDW of the traditional nonlinear fibers (highly-nonlinear fibers, dispersion-shifted fibers, and the highly-nonlinear dispersion-shifted fibers) is located around 1550 nm, and it can be tuned only in a very small wavelength range, which is determined by the fiber structure and the fabricated process. Since the PCF has flexible cladding and core structures, the ZDW can be tuned to nearly "any" wavelength in the transparent window. It means that the optical parametric components can be generated from visible to the infrared band flexibly.

Fiber-based optical parametric generation, also named as modulation instability, refers to the phenomenon that signals from the spontaneous emission are amplified by the pump wave. The shape and the distribution of the optical parametric spectral components are closely related to the nonlinear effect of FWM. Marhic et al. [3] reported a widely tunable optical parametric generation with the anti-Stokes spectral components around 1350 nm in a highly-nonlinear dispersion-shifted fiber pumped in the normal dispersion wavelength regime by a tunable diode laser. Wong et al. demonstrated the evolution of the optical parametric spectral components versus the pump wavelength by using a quasi-continuous wave laser source. When the PCF is pumped near the ZDW, a wide-band parametric spectrum can be formed. When the PCF is pumped in the normal dispersion wavelength regime, two narrow parametric bands can be generated far from the pump [4]. Harvey et al. demonstrated that the parametric spectral components can be tuned largely by adjusting the polarization state of the pump in a birefringent PCF [5].

It is clear that the FWM optical parametric spectral components can be easily generated by pumping the PCF near the ZDW. Andersen et al. obtained a widely tuned optical parametric generation in a PCF with two zero-dispersion wavelengths (TZDWs) by using a Ti: sapphire laser as the pump source [6]. Tuan et al. [7] theoretically predicted that largely tuned parametric components can be generated in a PCF with four ZDWs. The largely tuned optical parametric generation has a lot of applications, such as new wavelength laser source, large span wavelength converter, and entangled pair-photon source. In this section, we would like to design and fabricate a PCF with TZDW for the realization of large span optical parametric generation.

The PCF structure is designed by using the multipole method. In order to ensure the precision during the fabrication, hexagonal cladding structures are adopted. The fused silica is used as the substrate material. The designed cross-sectional structure of the PCF is shown in Fig. 2.1. The pitch of the air holes is 1.15 μm, the diameter of the air holes is 0.75 μm, and the diameter of the fiber core is 1.55 μm. The effective refractive index and the effective mode area at different wavelength can be calculated by using the multipole method. Based on the relation between the refractive index and the dispersion, the group velocity dispersion of the fundamental mode can be obtained, which is shown in Fig. 2.2. The two ZDWs are located at 701 and 1115 nm, respectively.

When a single-longitudinal-mode laser is used as the pump source, the frequencies and the propagation constants of the two pump photons are the same.

$$\omega_1 = \omega_2 = \omega_p \tag{2.4}$$

$$\beta_1 = \beta_2 = \beta_p \tag{2.5}$$

Fig. 2.1 The cross section of the designed PCF

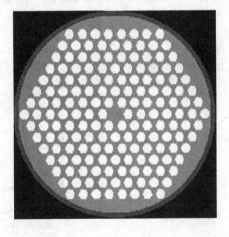

Fig. 2.2 The group velocity dispersion for the designed PCF with TZDW

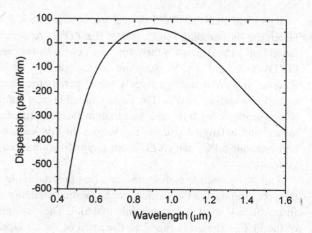

The phase-matching condition for the single pump case can be expressed as follows:

$$2\omega_p = \omega_s + \omega_i \tag{2.6}$$

$$\Delta k = \Delta \beta = \beta_s + \beta_i - 2\beta_p = 0 \tag{2.7}$$

The propagation constant can be calculated by the equation of $\beta = n\omega/c$. The phase-matching condition can be predicted by a combination of the Eqs. (2.6) and (2.7), which is shown in Fig. 2.3 plotted by pump wavelength on the horizontal axis and the wavelengths of the signal and idler on the vertical axis. In the figure, the blue line represents the signal and the red line represents the idler. When the pump is located at 780 nm, the phase-matched signal and idler arise at the telecommunication band of 1550 nm and the visible band of 521 nm, respectively, which is denoted by the purple line in the figure.

Fig. 2.3 The phase-matching profile for the designed PCF with TZDW

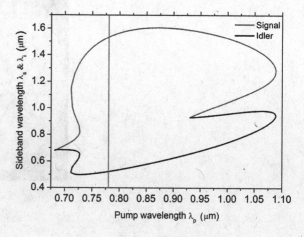

The designed PCF is featured with a small core, which can provide a high nonlinearity. The microstructure of the designed PCF introduces a lot of challenges to the technical fabrication. The group of Prof. Jinyan Li in Huazhong University of Science and Technology fabricated this fiber for us successfully by overcoming lots of difficulties. The SEM image of the cross section of the fabricated PCF is shown in Fig. 2.4. According to the fiber parameters extracted from the SEM image, the dispersion property is calculated, and the TZDWs are located at 723 and 1363 nm, respectively, as shown in Fig. 2.5. The nonlinear coefficient is calculated to be $160 \text{ km}^{-1} \text{ W}^{-1}$ at 800 nm.

Fig. 2.4 The SEM image of the cross section of the fabricated PCF with TZDW

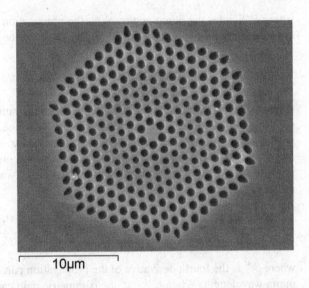

Fig. 2.5 The calculated dispersion for the fabricated PCF

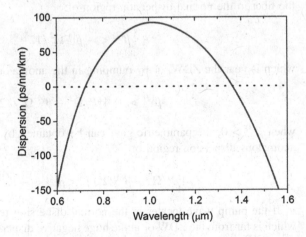

2.2.2   The Relationship Between the Optical Parametric Spectrum and the Pump Wavelength

In order to accurately simulate the optical parametric generation, the contribution of the nonlinear effect to the phase matching should be considered, and the phase-matching condition equation should be modified accordingly. For the single pump case, the revised phase-matching condition can be expressed as follows:

$$\kappa = \Delta\beta + 2\gamma P = \beta_s + \beta_i - 2\beta_p + 2\gamma P = 0 \tag{2.8}$$

where P denotes the pump power. With the propagation constants β_s and β_i Taylor expanded around the pump frequency, Eq. (2.9) can be written as follows:

$$\kappa = \Omega^2\beta_2 + \frac{1}{12}\Omega^4\beta_4 + \frac{1}{360}\Omega^6\beta_6 + \cdots + 2\gamma P = 0 \tag{2.9}$$

where Ω denotes the frequency detuning between the pump and the signal. It can be seen that only the even-order dispersion and the nonlinearity have influence on the phase-matching condition. When the linear phase mismatch is located in the region of $-4\gamma P < \Delta\beta < 0$, a nonzero parametric gain can be obtained. It requires that the group velocity dispersion parameter $\beta^{(2)}$ at the pump wavelength is approximately in the range of:

$$-\frac{4\gamma P}{\Omega^2} - \frac{\beta^{(4)}\Omega^2}{12} < \beta^{(2)} < -\frac{\beta^{(4)}\Omega^2}{12} \tag{2.10}$$

where $\beta^{(4)}$ is the fourth derivative of the propagation parameter β with respect to the pump wavelength. When $\beta^{(4)} < 0$, the parametric gain can be obtained by pumping the fiber in the normal dispersion region of:

$$\beta^{(2)} < -\beta^{(4)}\Omega^2/12 \tag{2.11}$$

which is near the ZDW, or by pumping in the anomalous dispersion region of:

$$\beta^{(2)} > -4\gamma P/\Omega^2 - \beta^{(4)}\Omega^2/12 \tag{2.12}$$

when $\beta^{(4)} > 0$, the parametric gain can be obtained by pumping the fiber in the anomalous dispersion region of:

$$-4\gamma P/\Omega^2 - \beta^{(4)}\Omega^2/12 < \beta^{(2)} < -\beta^{(4)}\Omega^2/12 \tag{2.13}$$

If the pump wave locates in the normal dispersion region of $\beta^{(2)} > |\beta^{(4)}\Omega^2/12|$, which is far from the ZDW or in the huge negative dispersion region of $\beta^{(2)} < -4\gamma P/\Omega^2 - |\beta^{(4)}\Omega^2/12|$, the parametric spectral components cannot be observed [8].

Fig. 2.6 The phase-matching contour for the nonlinear mismatch term of $\gamma P = 0$. Reprinted from Ref. [8], copyright © 2013, with permission from Elsevier

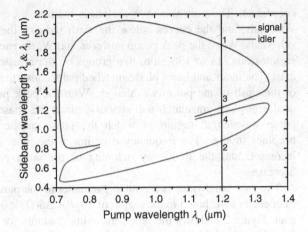

The phase-matching contour for the fundamental mode of the PCF with nonlinear mismatch term $\gamma P = 0$ is shown in Fig. 2.6. The range of the pump wavelength, in which phase-matched signal and idler pair exist, is from 710 to 1346 nm, beginning at the normal dispersion regime near the first ZDW of 723 nm, and ending at the anomalous dispersion regime close to the second ZDW of 1363 nm. For each pump wave in the regions from 710 to 832 nm and from 1120 to 1346 nm, two groups of phase-matched signal and idler pairs exist. For example, for the pump wavelength of 1200 nm, the outer pair of the phase-matched points is marked with 1 and 2, and the inner pair of the phase-matched points is marked with 3 and 4, as shown in the vertical solid line on Fig. 2.6. For each pump wave in the region from 832 to 1120 nm, only one group of phase-matched signal and idler pair exists, and the signal and the idler bands have a large interval. Figure 2.7 shows the evolution of the phase-matched sidebands versus the pump wavelength with different pump power. The black slash shows the location of the pump wavelength in the vertical

Fig. 2.7 The phase-matching contours for different pump power. Reprinted from Ref. [8], copyright © 2013, with permission from Elsevier

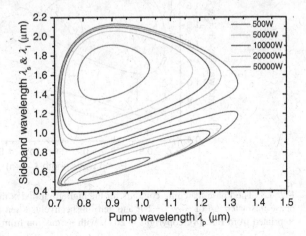

coordinate. The curves above the slash indicate the wavelengths of the signal sidebands, and the curves below the slash indicate the wavelengths of the idler sidebands. When the peak pump power is 500 W, for each pump wavelength in the region from 711 to 1330 nm, two groups of phase-matched signal and idler pairs exist. The inner and outer phase-matched pairs constitute a ring shape on each side of the slash of the pump wavelength. With the peak pump power increased, the nonlinear phase mismatch will seriously affect the phase-matching condition. The pump wavelength region in which the phase-matched waves pair can appear becomes smaller. The frequency detuning of the inner pair from the pump wave increases, and the frequency detuning of the outer pair from the pump wave decreases.

The gain bandwidth is a critical parameter to the parametric amplification. The frequencies distributed in the region of $-4\gamma P < \Delta\beta(\Omega) < 0$ will experience a nonzero gain. Figure 2.8 shows the phase-matching contours for the fundamental mode of the PCF when the linear phase mismatch $\Delta\beta$ equals to $-4\gamma P$, $-2\gamma P$ and zero, respectively, and the peak pump power P is 20,000 W. For a given pump wavelength, the gain band covers the wavelengths that satisfy the condition of $-4\gamma P < \Delta\beta(\Omega) < 0$. The phase-matching curve with $\Delta\beta = -2\gamma P$ indicates the signals and idlers with the maximal gain. Based on the profiles of the gain spectrum, the pump wavelengths are divided into six regions. In the region 1, for a given pump wave, there are four separated gain bands, corresponding to the signal and idler gain bands of the inner and outer pairs, respectively. For example, with the pump wavelength of 800 nm, the four gain bands are a, b, c, and d, respectively, which are marked in the Fig. 2.10. In the region 2, for a given pump, the signal and idler bands of the inner pair are connected to each other, and a broad band is formed near the pump wavelength except for two separated signal and idler bands of the outer

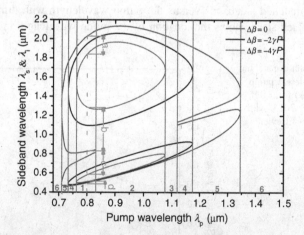

Fig. 2.8 The Phase-matching contours for the high group-index mode of the PCF with the linear phase mismatch $\Delta\beta$ of $-4\gamma P$, $-2\gamma P$ and zero, respectively, when the pump power is 20,000 W. Reprinted from Ref. [8], copyright © 2013, with permission from Elsevier

pair. In the region 3, for a given pump, the four gain bands are all connected to each other and form a superbroad gain band similar to a supercontinuum spectrum. In the region 4, for a given pump, the idler or signal of the outer and inner gain bands are combined together. Either the signal or the idler band includes two gain peaks since two fulfilled phase-matching wavelengths are existed. In the region 5, the fulfilled phase-matched wavelengths do not exist, and the parametric gain is very weak. The region 6 belongs to zero gain region, and no parametric gain exists. It can be predicted that, according to band gain contours, various parametric gain shapes can be obtained by adjusting the pump wavelength.

2.2.3 The Optical Parametric Generation in a PCF with TZDW

The optical parametric spectra are measured for the PCF with TZDW. The experimental setup is shown in Fig. 2.9. Experimentally, a Ti: sapphire pulse laser can emit a pulse train with the full width at half maximum (FWHM) of 130 fs, at the repetition rate of 76 MHz. The pump pulse train is coupled into 1.0-m PCF mentioned above through a $40 \times$ microscope objective lens with the numerical aperture of 0.65. The central wavelength of the pump wave is set to be 800 nm. The pump power can be adjusted by a neutral-density filter wheel. The light emitting from the fiber is collimated by a $40 \times$ microscope objective lens and then sent to two optical fiber spectrometers (Avaspec-2048-2 and Avaspec-NIR-256-2.5) with the measurement scopes from 200 to 1100 nm and from 900 to 2500 nm.

Since the signal band distributing in the wavelength region of 1000–2200 nm, a suitable signal source is not accessible. We used amplified spontaneous emission (ASE) from the pump laser as the seed source and inferred the FOPA gain spectrum from the measurement of the output ASE spectrum. The experimentally observed spectra for the peak pump power of 20,000 W and the pump wavelengths of 760, 800 and 815 nm are shown in Fig. 2.10. The idler bands are shown in Fig. 2.10a, c, and e, in which two gain bands in visible region can be clearly seen. When the pump is operated at 800 nm, the region of 550–730 nm corresponds to the idler wave of the inner pair (IWIP) of the sideband, and the region of 410–501 nm corresponds to the idler wave of the outer pair (IWOP) of the sideband. When the pump wavelength increases, both the IWIP and IWOP move to the longer

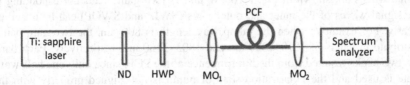

Fig. 2.9 The experimental setup

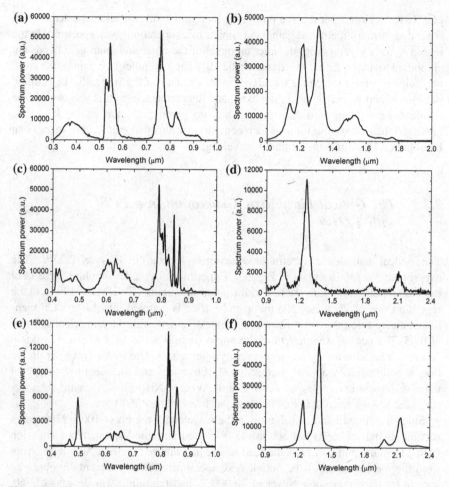

Fig. 2.10 The observed output spectra of the idler waves, with the pump wavelengths of **a** 0.76 μm, **c** 0.8 μm, and **e** 0.815 μm, respectively. The observed output spectra of the signal waves, with the pump wavelengths of **b** 0.76 μm, **d** 0.8 μm, and **f** 0.815 μm, respectively. The peak pump power is set at $P = 20,000$ W. Reprinted from Ref. [8], copyright © 2013, with permission from Elsevier

wavelength, which is shown in Fig. 2.11. In particular, when the pump wavelength is 760 nm, the idler wave extends down to the ultraviolet region of 300 nm. The signal waves are shown in Fig. 2.10b, d, and f. Two gain bands corresponding to the signal waves of the inner and outer pairs (SWIP and SWOP) can be observed clearly. For example, when the pump wavelength is 800 nm, the two gain bands distribute in bands of 1010–1320 nm and 1805–2160 nm, respectively. Each band has two peaks resulted from the birefringence of the PCF, since only one half-wave plate is used and the polarization state of pump is not aligned properly with the principle axis of the PCF. Theoretically, a half-wave plate and two quarter-wave

Fig. 2.11 Sideband wavelength of the two pairs of FWM versus the pump wavelength, the peak pump power is set at 20,000 W. Reprinted from Ref. [8], copyright © 2013, with permission from Elsevier

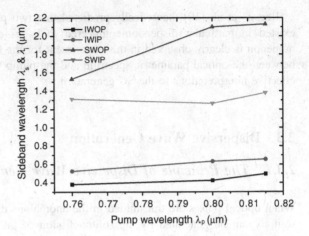

plates should be used simultaneously so that all the polarization states can be reached. The evolution of the signal band versus the pump wavelength is also shown in Fig. 2.11. When the pump wavelength is 815 nm, the signal band extends to the mid-infrared region of 2190 nm. The two pairs of gain bands are approximately in agreement with the simulation results in Fig. 2.7.

In order to further study the evolution of the two pairs FWM gain bands versus the propagation distance, we simulate the $sech^2$ pulses with peak power of 20,000 W and 130 fs FWHM propagating in 1-m PCF using the split-step Fourier method to solve the generalized nonlinear Schrödinger equation [9]. The spectral evolution is shown in Fig. 2.12. It is clear that two pairs of sideband are generated at the propagation distance about 3 mm. As shown in Fig. 2.12, the outer pair of FWM gain bands is composed of A and B, and the inner pair is composed of C and D. The wavelengths of them remain relatively fixed with the increasing of the propagation distance to 1 m.

Fig. 2.12 Simulated spectral evolution of a pulse launched at 800 nm with the peak pump power of 20,000 W. Reprinted from Ref. [8], copyright © 2013, with permission from Elsevier

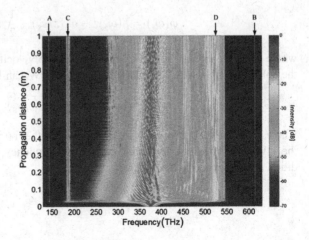

Reeves et al. [10] theoretically predicted that two pairs of gain bands can be existed in particular dispersion-engineered PCFs as early as 2003. This phenomenon is clearly observed in this experiment for the first time. The relationship between the optical parametric spectrum and the pump wavelength provides more effective interpretations to the SC generation.

2.3 Dispersive Wave Generation

2.3.1 The Principle of Dispersive Wave Generation

When optical pulses are transmitted in the anomalous dispersion regime, multiple solitons can be generated by the soliton fission of the pump pulse. The soliton frequency can be continuously shifted to the longer wavelength region by the intra-pulse Raman scattering. When the solitons are perturbed by third-order and higher order dispersion, the energy can be transferred to the dispersive waves (DWs) located in the normal dispersion regime [11]. The DW is also named as Cherenkov radiation (CR) or nonsolitonic radiation (NSR). The SC can be broadened to the shorter and longer wavelength regions by the DW [12–16]. Tunable DW in the wavelength region from 485 to 600 nm has been reported in a PCF [17, 18]. By using a variety of PCFs with different group velocity dispersion, the DW wavelength can even be extended to UV region [19–21]. Based on a PCF with TZDW, large amplitude DW near 1600 nm have been generated in the longer normal dispersion wavelength region, which is beyond the second ZDW [22]. The DW at the wavelength of 1720 nm has been obtained experimentally by Schreiber et al. [14].

The DW generation requires that the phase-matching condition of the DW propagates with the same phase velocity as that of the soliton should be satisfied. The phase shift of soliton can be expressed as follows:

$$\phi(\omega_s) = \beta(\omega_s)z - \omega_s(z/v_g) + \frac{1}{2}\gamma P_s z \qquad (2.14)$$

where ω_s is the soliton frequency, v_g is the group velocity of the soliton, and P_s is the soliton peak power. The phase shift of the DW can be expressed as follows:

$$\phi(\omega_d) = \beta(\omega_d)z - \omega_d(z/v_g) \qquad (2.15)$$

where ω_d is the DW frequency. After expanding the mode-propagation constant of the DW in a Taylor series about the soliton frequency, the phase-matching condition can be expressed as follows:

$$\sum_{n \geq 2} \frac{(\omega_d - \omega_s)^n}{n!} \beta_n = \frac{\gamma P_s}{2} \tag{2.16}$$

In the qualitative analysis, the value of ω_{CR} is mainly affected by the first few terms of the left part of Eq. (2.16). When only the first two terms on the left part are taken into account, the Eq. (2.16) can be written simply as follows:

$$\frac{(\omega_d - \omega_s)^2}{2} \beta_2 + \frac{(\omega_d - \omega_s)^3}{6} \beta_3 = \frac{\gamma P_s}{2} \tag{2.17}$$

Physically, the right part is always positive. The soliton can be formed at the anomalous dispersion regime ($\beta_2 < 0$). When the third-order dispersion (TOD) is positive ($\beta_3 > 0$), the value of ω_{CR} is larger than that of ω_s, and the CR shifts to shorter wavelength regions. When the TOD is negative $\beta_3 < 0$, the value of ω_{CR} is smaller than that of ω_s, and the CR shifts to the longer wavelength region.

2.3.2 Tunable DWs in the Visible Region

In the experiment, a mode-locked Ti: sapphire femtosecond pulsed laser is used as the pump source. A highly-nonlinear PCF with TZDW is used as the nonlinear medium. The calculated dispersion parameter β_2 and the TOD parameter β_3 versus the wavelength are shown in Fig. 2.13. In the vicinity of 1025 nm, β_3 moves from positive to negative.

When the pump wave is operated at 800 nm with an average power of 70 mW, the spectrum obtained from the optical fiber spectrometer is shown in Fig. 2.14. The

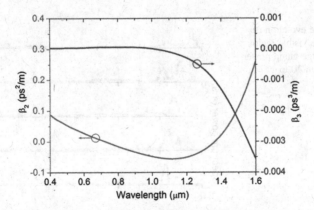

Fig. 2.13 The calculated dispersion parameter β_2 and third-order dispersion parameter β_3 for the fundamental mode of the PCF. The inset shows the scanning electron microscope (SEM) image of the cross section of the PCF

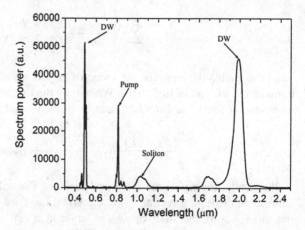

Fig. 2.14 The experimentally measured spectrum for the pump operating at 800 nm with an average power of 70 mW

soliton is firstly formed near the pump wavelength, where $\beta_3 > 0$, and the DW is generated at 498 nm based on the phase-matching condition between the soliton and DW. Then, the soliton shifts toward the longer wavelength region because of the RIFS. When the soliton approaches the second ZDW, where $\beta_3 < 0$, the DW is generated at 1986 nm. The RIFS is canceled by the spectral recoil from the generation of the DW at 1986 nm, and the soliton is stabilized in the wavelength region close to the second ZDW [23].

The wavelength of the DW in the visible region is studied by adjusting the average pump power. As shown in Fig. 2.15, with the average pump power increasing from 70 to 400 mW, the wavelength of DW can be tuned from 498 to 425 nm. The conversion efficiency is calculated, which is shown by the black line in Fig. 2.16. The conversion efficiency of the visible DW increases monotonously with the average pump power. It reaches 42 % at the average pump power of 400 mW [24].

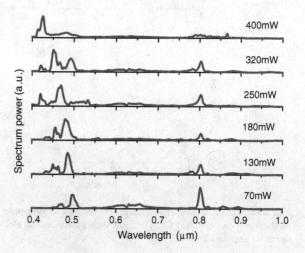

Fig. 2.15 The evolution of the DW in the visible region with the average pump power increasing from 70 to 400 mW

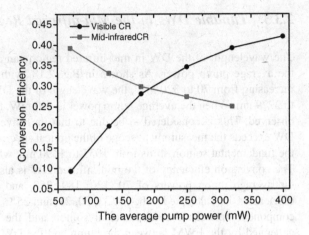

Fig. 2.16 The conversion efficiencies of the visible and mid-infrared DWs versus the average pump power

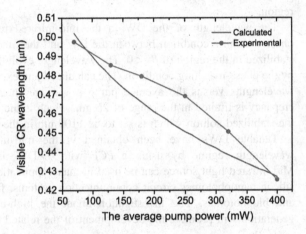

Fig. 2.17 The calculated and experimentally measured DW wavelength in the visible as a function of the average pump power

The wavelength of the DW is governed by the phase-matching condition between the DW and the soliton formed near the pump wavelength, which is in the region of $\beta_3 > 0$. Based on the phase-matching condition, the DW wavelength is predicted. In the simulation, dispersive effects up to 10th order are included. The calculated and the experimentally measured DW wavelengths versus the average pump power are shown in Fig. 2.17. The discrepancy is limited in the range of 18 nm, which is induced by the wavelength of the soliton which is set to be 800 nm in the simulation. The trend of the predicted DW wavelength evolution is a little different with that obtained in the experiment. This could be induced by the accuracy of the simulated dispersion terms. Except for the DW generation, FWM process can also happen. For each pump wavelength, two groups of phase-matched signal and idler pairs exist. The spectral component between the pump wave and the DW is considered to be generated by the FWM between the pump wave and the red-shifted fundamental soliton. The idler of the outer group of the FWM distribute in the vicinity of the DW.

2.3.3 Tunable DWs in the Mid-infrared Region

The wavelength of the DW in mid-infrared region can also be tuned by adjusting the average pump power. As shown in Fig. 2.18, with the average pump power increasing from 70 to 320 mW, the wavelength of the DW can be tuned from 1986 to 2279 nm. When the average pump power is 400 mW, the mid-infrared DW is not observed. This is considered to be due to that the wavelength of the mid-infrared DW exceeds the measurement scope of the optical spectrometer. The wavelength of the fundamental soliton shifts from 1035 to 1190 nm, where β_3 is always negative. The conversion efficiency of the mid-infrared DW is also shown in Fig. 2.16. For the average pump powers of 70, 130, 180, 250, and 320 mW, the conversion efficiencies are calculated to be 39, 33, 30, 27, and 25 %, respectively. The spectral component between the fundamental soliton and the DW is considered to be generated by the FWM between the pump and the DW in the visible wavelength region.

The wavelength of the DW in the mid-infrared region is governed by the phase-matching condition between the DW and the fundamental soliton, which is stabilized in the region of $\beta_3 < 0$. The wavelength evolution of the DW is predicted by the phase-matching condition. The calculated and experimentally measured DW wavelengths versus the average pump power are shown in Fig. 2.19. The discrepancy is limited in the range of 26 nm, which is induced by the wavelength of the stabilized soliton which is set to be 1100 nm in the simulation [25].

Tunable DWs have been obtained in the mid-infrared normal dispersion wavelength regime by using a PCF with TZDW as the nonlinear medium. Mid-infrared light source can be used in many applications, such as the integrated silicon nanophotonics circuit communication systems, free-space communication, and biophotonics. The investigation about the high-efficient mid-infrared DW generation will promote the development of the related areas.

Fig. 2.18 The evolution of the DW in the mid-infrared region with the average pump power increasing from 70 to 320 mW

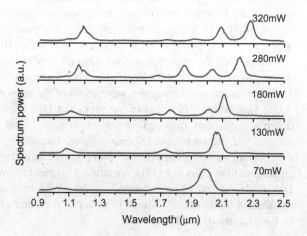

Fig. 2.19 The calculated and experimentally measured mid-infrared DW wavelengths as a function of the average pump power

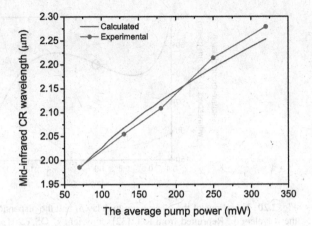

2.4 Ultraviolet Generation Based on Cross-Phase Modulation in PCF with TZDW

Supercontinuum (SC) spanning from the visible to the near-infrared region has been widely investigated and utilized in many fields [26]. Nonetheless, recently many efforts have been devoted to extend the SC to mid-infrared and ultraviolet (UV) regions [27, 28]. Light source in the UV region is urgently needed in the applications of fluorescence microscopy, spectroscopy, and biomedical photonics. DW generation and four-wave mixing (FWM) are effective ways to extend the SC to short wavelength region. However, both of them require the phase-matching condition, which is difficult to fulfill in the UV wavelength region. UV SC had been generated by pumping a PCF with a picosecond microchip laser at 355 nm. However, due to the large normal dispersion at UV region and the effect of Raman-induced frequency shift, the spectrum is mainly extends to longer wavelength regions [29].

Except for the DW generation and FWM, cross-phase modulation (XPM) is also an effective way to extend the SC to shorter wavelength regions. In an experiment, dual-frequency (1064 and 532 nm) pulsed pump waves are simultaneously coupled into a photonic crystal fiber (PCF). The Raman-induced frequency shift of the pulse at 532 nm is inhibited by the XPM induced by the pulse at 1064 nm, and the SC extends to shorter wavelength region of 350 nm [30]. SC has been extended to short wavelength region of 300 nm by the effect of XPM between the anti-Stokes signal at 550 nm and the pump pulse at 830 nm [31].

When the PCF with TZDW is pumped by a Ti: Sapphire pulse at 800 nm, one of the anti-Stokes signals appears in the region from 400 to 550 nm. The calculated relative group delay β_1 with respect to the pump wavelength of 800 nm and the dispersion parameter β_2 versus the wavelength are shown in Fig. 2.20. It is predicted that XPM between the anti-Stokes signal and the Raman soliton generated by the pump can be occurred.

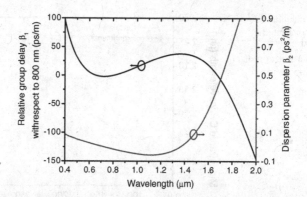

Fig. 2.20 Variations of the relative group delay β_1 and the dispersion parameter β_2 with respect to the wavelength. Reprinted from Ref. [32], copyright © OSA 2013, with permission from OSA

The experimental spectra for the average pump power of 200, 300, and 400 mW are shown in Fig. 2.21 For the input average pump power of 200 mW, the inner and outer anti-Stokes signals are observed at 550–750 nm and 400–550 nm, respectively. This is consistent with the simulation results about the two pairs of FWM. According to the relative group delay in Fig. 2.20, the outer anti-Stokes signal has a lower group velocity than the pump soliton. As the Raman-induced frequency shift, the pump soliton shifts to longer wavelengths, and the group velocity slows down. Then, the soliton overlaps with the outer anti-Stokes signal temporally. The anti-Stokes signal is affected by the XPM induced by the pump soliton. The SC is extended to the UV region of 200–400 nm. As the input power is increased, the frequency component in the UV region of 200–400 nm tends to become flat [32]. The experimental scene is shown in Fig. 2.22. Figure 2.23 shows the optical spot of the fundamental mode at the output.

Fig. 2.21 The experimental spectra for the average pump powers of 200, 300, and 400 mW. Reprinted from Ref. [32], copyright © OSA 2013, with permission from OSA

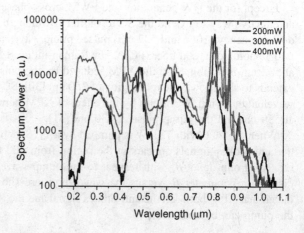

Fig. 2.22 The experimental scene of the XPM

Fig. 2.23 The optical spot of the fundamental mode at the output

2.5 Conclusion

In this chapter, we first investigate the relationship between the FWM parametric spectrum and the dispersion, nonlinear coefficient, and pump wavelength. A PCF with TZDW is carefully designed and fabricated. The parametric spectra in this PCF are measured by using a Ti: sapphire laser as the pump. Two pairs of large span parametric sidebands are observed, and the sideband in mid-infrared region can be extended to 2190 nm.

Second, the DW generation is investigated in this PCF. In the experiment, visible and mid-infrared DWs can be generated simultaneously by pumping in the anomalous dispersion regime between the TZDWs. The visible DW can be generated from 498 to 425 nm, and the mid-infrared DW can be tuned from 1986 to 2279 nm.

Third, the XPM effect is investigated in this PCF. The FWM and DW generation effects can emit an anti-Stokes signal in the blue–violet region. Due to the XPM between the anti-Stokes signal and the Raman soliton, the SC can extend to the ultraviolet region of from 200 to 400 nm.

References

1. T.P. White, B.T. Kuhlmey, R.C. McPhedran, D. Maystre, G. Renversez, C. Martijn de Sterke, L.C. Botten, Multipole method for microstructured optical fibers. I. Formulation. J. Opt. Soc. Am. B **19**(10), 2322–2330 (2002)
2. B.T. Kuhlmey, T.P. White, G. Renversez, D. Maystre, L.C. Botten, C. Martijn de Sterke, R.C. McPhedran, Multipole method for microstructured optical fibers. II. Implementation and results. J. Opt. Soc. Am. B **19**(10), 2331–2340 (2002)
3. M.E. Marhic, K.K.Y. Wong, L.G. Kazovsky, Wide-band tuning of the gain spectra of one-pump fiber optical parametric amplifiers. IEEE J. Sel. Top. Quant. Electron **10**(5), 1133–1141 (2004)
4. G. Wong, A. Chen, S.G. Murdoch, R. Leonhardt, J.D. Harvey, N.Y. Joly, J.C. Knight, W.J. Wadsworth, P.J. Russell, Continuous-wave tunable optical parametric generation in a photonic-crystal fiber. J. Opt. Soc. Am. B **22**(11), 2505–2511 (2005)
5. J.D. Harvey, R. Leonhardt, S. Coen, G.K.L. Wong, J.C. Knight, W.J. Wadsworth, P. Russell, Scalar modulation instability in the normal dispersion regime by use of a photonic crystal fiber. Opt. Let. **28**(22), 2225–2227 (2003)
6. T.V. Andersen, K.M. Hilligsøe, C.K. Nielsen, J. Thøgersen, K.P. Hansen, S.R. Keiding, J.L. Larsen, Continuous-wave wavelength conversion in a photonic crystal fiber with two zero-dispersion wavelengths. Opt. Express **12**(17), 4113–4122 (2004)
7. T.H. Tuan, T. Cheng, K. Asano, Z. Duan, W. Gao, D. Deng, T. Suzuki, Y. Ohishi, Optical parametric gain and bandwidth in highly nonlinear tellurite hybrid microstructured optical fiber with four zero-dispersion wavelengths. Opt. Express **21**(17), 20303–20312 (2013)
8. L. Zhang, S.-G. Yang, Y. Han, H.-W. Chen, M.-H. Chen, S.-Z. Xie, Optical parametric generation with two pairs of gain bands based on a photonic crystal fiber. Opt. Commun. **300**, 22–26 (2013)
9. G.P. Agrawal, *Nonlinear Fiber Optics* (Springer, Berlin Heidelberg, 2000)
10. W.H. Reeves, D.V. Skryabin, F. Biancalana, J.C. Knight, P.J. Russell, F.G. Omenetto, A. Efimov, A.J. Taylor, Transformation and control of ultrashort pulses in dispersion-engineered photonic crystal fibres. Nature **424**(6948), 511–515 (2003)
11. N. Akhmediev, M. Karlsson, Cherenkov radiation emitted by solitons in optical fibers. Phys. Rev. A **51**(3), 2602–2607 (1995)
12. B.H. Chapman, J.C. Travers, S.V. Popov, A. Mussot, A. Kudlinski, Long wavelength extension of CW-pumped supercontinuum through soliton-dispersive wave interactions. Opt. Express **18**(24), 24729–24734 (2010)
13. D.R. Austin, C.M. de Sterke, B.J. Eggleton, T.G. Brown, Dispersive wave blue-shift in supercontinuum generation. Opt. Express **14**(25), 11997–12007 (2006)
14. T. Schreiber, T. Andersen, D. Schimpf, J. Limpert, A. Tünnermann, Supercontinuum generation by femtosecond single and dual wavelength pumping in photonic crystal fibers with two zero dispersion wavelengths. Opt. Express **13**(23), 9556–9569 (2005)
15. V. Husakou, J. Herrmann, Supercontinuum generation in photonic crystal fibers made from highly nonlinear glasses. Appl. Phys. B Lasers Opt. **77**(2–3), 227–234 (2003)
16. M. Frosz, P. Falk, O. Bang, The role of the second zero-dispersion wavelength in generation of supercontinua and bright-bright soliton-pairs across the zero-dispersion wavelength. Opt. Express **13**(16), 6181–6192 (2005)
17. G. Chang, L.J. Chen, F.X. Kärtner, Highly efficient Cherenkov radiation in photonic crystal fibers for broadband visible wavelength generation. Opt. Let. **35**(14), 2361–2363 (2010)
18. J. Yuan, X. Sang, C. Yu, Y. Han, G. Zhou, S. Li, L. Hou, Highly efficient and broadband Cherenkov radiation at the visible wavelength in the fundamental mode of photonic crystal fiber. IEEE Photon. Technol. Lett. **23**(12), 786–788 (2011)
19. H. Tu, S.A. Boppart, Optical frequency up-conversion by supercontinuum-free widely unable fiber optic Cherenkov radiation. Opt. Express **17**(12), 9858–9872 (2009)

20. H. Tu, S.A. Boppart, Ultraviolet-visible non-supercontinuum ultrafast source enabled by switching single silicon strand-like photonic crystal fibers. Opt. Express **17**(20), 17983–17988 (2009)
21. S.P. Stark, A. Podlipensky, N.Y. Joly, P.J. Russell, Ultraviolet-enhanced supercontinuum generation in tapered photonic crystal fiber. J. Opt. Soc. Am. B **27**(3), 592–598 (2010)
22. M. Erkintalo, G. Genty, J.M. Dudley, Giant dispersive wave generation through soliton collision. Opt. Lett. **35**(5), 658–660 (2010)
23. D.V. Skryabin, F. Luan, J.C. Knight, P.J. Russell, Soliton self-frequency shift cancellation in photonic crystal fibers. Science **301**(5640), 1705–1708 (2003)
24. L. Zhang, S.-G. Yang, Y. Han, H–.W. Chen, M–.H. Chen, S–.Z. Xie, Simultaneous generation of tunable giant dispersive waves in the visible and mid-infrared regions based on photonic crystal fibers. J. Opt. **15**(7), 075201 (2013)
25. L. Zhang, S. Yang, H. Chen, M. Chen, S. Xie, in Wavelength-tunable red-shift Cherenkov radiation in photonic crystal fibers for mid-infrared wavelength generation. *18th Opto-Electronics and Communications Conference, OECC*: 2013, WS4–4
26. J.M. Dudley, G. Genty, S. Coen, Supercontinuum generation in photonic crystal fiber. Rev. Mod. Phys. **78**(4), 1135 (2006)
27. J. Yuan, X. Sang, C. Yu, K. Wang, B. Yan, X. Shen, Y. Han, G. Zhou, S. Li, L. Hou, Widely wavelength-tunable two-colored solitons and small spectral component for broadband mid-infrared wavelength generation in a highly birefringent photonic crystal fiber. IEEE Photon. Technol. Lett. **24**(8), 670–672 (2012)
28. S.P. Stark, J.C. Travers, P.S.J. Russell, Extreme supercontinuum generation to the deep UV. Opt. Lett. **37**(5), 770–772 (2012)
29. T. Sylvestre, A.R. Ragueh, M.W. Lee, B. Stiller, G. Fanjoux, B. Barviau, A. Mussot, A. Kudlinski, Black-light continuum generation in a silica-core photonic crystal fiber. Opt. Lett. **37**(2), 130–132 (2012)
30. P.A. Champert, V. Couderc, P. Leproux, S. Février, V. Tombelaine, L. Labonté, P. Roy, C. Froehly, P. Nérin, White-light supercontinuum generation in normally dispersive optical fiber using original multi-wavelength pumping system. Opt. Express **12**(19), 4366–4371 (2004)
31. Y. Han, L.-T. Hou, J.-H. Yuan, C.-M. Xia, G.-Y. Zhou, Ultraviolet continuum generation in the fundamental mode of photonic crystal fibers. Chin. Phys. Lett. **29**(1), 014201 (2012)
32. L. Zhang, S. Yang, H. Chen, M. Chen, S. Xie, in Ultraviolet-shift supercontinuum generation by cross-phase modulation in photonic crystal fiber. *Conference on Lasers and Electro-Optics: Laser Science to Photonic Applications, CLEO*: 2013, JW2A.13

Chapter 3
PCF-Based Optical Parametric Amplifier (OPA)

3.1 Introduction

The emergence of the optical amplification technology has accelerated the development of the optical communications. The lifestyle has been rapidly changed. Furthermore, the optical amplification technology has potential applications in the fields of high-power lasers, signal regeneration, biomedicine, and laser weapons. Recently, the research about the optical amplification technology has been growing rapidly.

The present optical amplifiers can be divided into the categories of doped fiber amplifier, semiconductor optical amplifier (SOA), fiber Raman amplifier (FRA), and fiber optical parametric amplifier (FOPA). Benefiting from the demands in the optical communications, the doped fiber amplifiers have been fully developed. The operating wavelength range is limited by the doped rare earth ions. Currently, they can be operated at the telecommunication band and the 1064-nm band. A gain larger than 30 dB can be provided. The SOA can be integrated with the diode lasers and can be operated in about 100-nm wavelength ranges near the 1300 and 1550 nm. A gain between 15 and 30 dB can be provided. The FRA utilizes the stimulated Raman scattering in optical fibers, which can amplify the signals in a wide band smaller than the pump frequency. It can be operated in the special wavelength bands where the doped fiber amplifier and SOA cannot work. The FRA requires a long fiber length. It has been used in the fields of communication and distributed sensing.

FOPA utilizes the FWM effect in fibers. The pump energy can be transferred to the signal, and a corresponding idler can also be generated. The bandwidth of the parametric gain can be enhanced by the increase of the pump power. The relationship can be expressed as

$$\Delta\omega \approx \left(\frac{\gamma P_0}{\beta_4}\right)^{1/4} \tag{3.1}$$

© Springer-Verlag Berlin Heidelberg 2016
L. Zhang, *Ultra-Broadly Tunable Light Sources Based on the Nonlinear Effects in Photonic Crystal Fibers*, Springer Theses,
DOI 10.1007/978-3-662-48360-2_3

where P_0 is the pump power, γ is the fiber nonlinear coefficient, and β_4 is the fourth-order dispersion coefficient [1]. In theory, a wide parametric gain band over 100 nm can be obtained by using a high-power pump. The bandwidth largely exceeds that of the EDFA and FRA. Furthermore, the FOPA can be potentially operated at any wavelength region, which is similar with the FRA. It makes sense to amplify the signals located at novel wavelength band, where the EDFA and SOA cannot be operated. In the FOPA systems, the operating wavelength region can cover a broadband by tuning the pump wavelength in several nanometers. Meanwhile, a high gain can be obtained from an FOPA, and the small signal gain can be expressed as follows:

$$G_s = 8.6\gamma P_0 L - 6 \tag{3.2}$$

where L is the fiber length [1]. The parametric gain can be increased by using a long fiber or a high-power pump. Potentially, the gain is much larger than that provided by the EDFA, SOA, or FRA. The noise level is very small, which is similar with the EDFA. Usually, the noise level is about 3 dB. When two cascaded FOPAs are used, the noise level can even be reduced to close 0 dB.

The FOPA can also be used in wavelength conversion. The idler at new wavelength can be generated. When the signals are modulated, the modulation information can be embedded in the idler. Potentially, the communication systems in the ocean, land, and free space can be well jointed by this functionality. The FWM is an instant phenomenon. When the pump is intensity modulated, the signal and idler will be intensity modulated simultaneously. Based on this property, the FOPA can be used for signal shaping and signal regeneration. In the FOPA system, the signal and idler photons are generated simultaneously, so it can be used to generate entangled photons. Due to the advantages and the wide applications, the FOPAs attract a lot of attentions. The characteristics are enhanced, and the operating wavelength regions are continuously expanded.

In 1996, the parametric gain is measured in a single-mode fiber by using a 1550-nm CW laser as the pump [2]. Then, OPAs based on the HNLF and DSF-HNLF are developed. Their ZDWs are located near 1550 nm, where the well-developed lasers in C band can be selected as the pump source. Hansryd et al. [3] reported a high gain of 49 dB in a 500-m-long HNLF. In 2006, Torounidis et al. demonstrated a gain over 70 dB in a 200-m-long HNLF pumped by a 2.8-W CW laser. Meanwhile, the parametric gain bandwidth is continuously broadened [4]. In 1996, an FOPA with 80-nm gain bandwidth is reported [5]. Then, Torounidis et al. [6] demonstrated a gain bandwidth over 100 nm. Ho et al. [7] reported that a gain bandwidth over 200 nm can be obtained by combining the FWM effect and the stimulated Raman scattering.

Benefiting from the flexible dispersion property of the PCF, the FOPA can be operated at nearly any wavelength region in the transparent band of the fiber [8]. In 2009, Sylvestre et al. reported a PCF-based OPA operated in the wavelength region near 1064 nm. A Q-switched laser operated at 1064 nm with a pulse width of 450 ps and a repetition rate of 1 kHz was used as the pump. The small signal is from a PCF-based SC source. The parametric gain can be obtained in a bandwidth of 50 nm. A maximal gain of 35 dB was obtained [9]. In 2010, Lee et al. reported an fully fiber-integrated OPA by using a PCF as the gain medium. A DFB laser was used as the pump source. Its wavelength can be tuned from 1040 to 1080 nm. Another DFB laser was used as the signal source. A gain bandwidth of 30 nm was obtained, and a maximal gain over 30 dB was obtained [10]. In 2012, Mussot et al. obtained a large gain in a wide bandwidth over 20 THz in a PCF-based OPA. A maximal gain over 40 dB was achieved [11].

3.2   Parametric Generation in a PCF with ZDW Near 1.06 μm

The FWM has a strict requirement to the ZDW of the PCF. In Chap. 2, a PCF with TZDW was designed and fabricated. Two pairs of FWM parametric gain bands can be generated by pumping in the anomalous dispersion wavelength regime. Furthermore, tunable radiation can be generated in the ultraviolet, visible, and mid-infrared regions by the nonlinear effects of dispersive wave generation and cross-phase modulation. However, the required pump source Ti: sapphire laser is featured as huge dimension, high cost, and complexity for operation. It is hard to be used in practice. In this chapter, a dispersion-flattened PCF with the ZDW located near 1064 nm is designed and fabricated. In this wavelength region, the ytterbium-doped fiber laser can be used as the pump source. It is cheap and stable. The FOPA operated at 1064-nm wavelength region can be achieved.

The designed cross section of the PCF is shown in Fig. 3.1. The hexagonal air hole structure is used. The pitch between the air holes is 2.5 μm. The diameter of the air holes is 0.9 μm, and the diameter of the core is 4.1 μm. The effective refractive index for the fundamental mode of the PCF can be calculated by using the multipole method. Then, the group velocity dispersion can be calculated, which is illustrated in Fig. 3.2.

According to the designed structure, the FiberHome Telecommunication Technologies Co. Ltd. fabricated several PCF samples. The SEM image of the cross section of one of the PCFs is shown in Fig. 3.3. Based on the structural parameters extracted from the SEM image, the fiber properties are calculated by using the multipole method. The FWM parametric gain bands are measured in a PCF with the ZDW located at 1062 nm. The nonlinear coefficient of this fiber is calculated to be 15 $W^{-1}km^{-1}$, and the fiber loss is measured to be 25 dB/km by using the cutback method at 1064 nm.

The experimental schematic diagram for measurement of the FWM parametric gain sidebands is shown in Fig. 3.4. The pump source is a homemade tunable mode-locked ytterbium-doped fiber laser. A semiconductor saturable absorber mirror (SESAM) is served as the mode locker. The gain is provided by a 1-m-long ytterbium-doped fiber (YDF). A circulator with 35-dB isolation from port 3 to port 2 and from port 2 to port 1 acts as an isolator to ensure unidirectional propagation

Fig. 3.1 The cross section of the theoretically designed PCF with the ZDW located near 1064 nm

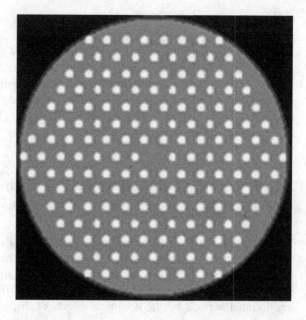

Fig. 3.2 The calculated group velocity dispersion for the fundamental mode of the designed PCF

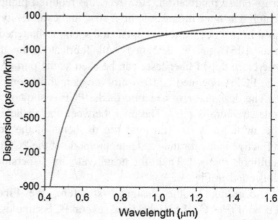

and suppress undesired reflections. A polarization controller (PC) is inserted in the cavity to adjust the polarization state of the light inside the cavity. A tunable band-pass filter (TBPF) is selected to tune the output wavelength. The wavelength can be tuned from 1033 to 1070 nm. The average output power is about 5 dBm. The electrical spectrum can be measured by a 10-GHz photodetector (PD), which is shown in Fig. 3.5. It can be seen clearly that the repetition rate is 25.4 MHz. The optical spectrum is measured by an optical spectrum analyzer (OSA). Figure 3.6 shows the optical spectrum for the output with a central wavelength of 1060.5 nm, and the 3-dB bandwidth is measured to be 2.7 nm. The full width at half maximum (FWHM) is measured to be 35 ps, which is shown by the solid line in Fig. 3.7. The

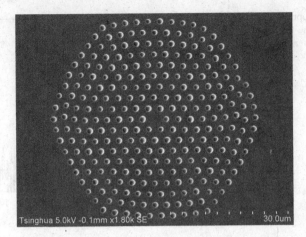

Fig. 3.3 The SEM image of the cross section of the fabricated PCF with the ZDW located near 1064 nm

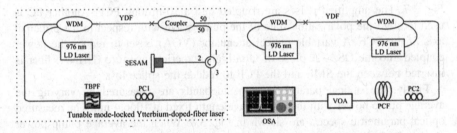

Fig. 3.4 The experimental schematic diagram of the FWM parametric generation

Fig. 3.5 The electrical spectrum of the generated pulse

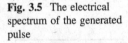
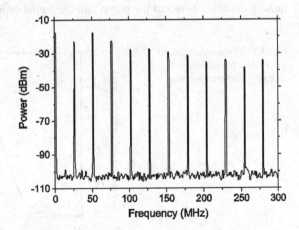

Fig. 3.6 The optical spectrum for the generated pulse train centered at 1060.5 nm

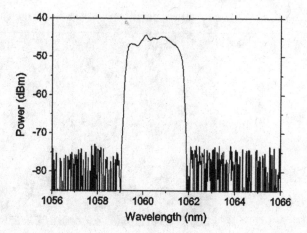

output pulses are amplified by a bidirectional core-pumped fiber amplifier using a 1-m-long YDF. After the amplifier, the average pump power can be amplified to 19 dBm, and the FWHM is measured to be 21 ps, which is shown by the dashed line in Fig. 3.7. The amplified pulses are coupled into a 10-m-long PCF. Another PC is used to align the polarization state of the pump wave with respect to the principal axis of the PCF. A variable optical attenuator (VOA) is set to reduce the power coupled into the OSA. A piece of ultra-high numerical aperture (UHNA) fiber is inserted between the SMF and the PCF to reduce the splice loss.

Firstly, the optical parametric gain sidebands are measured by varying the average pump power with the pump wavelength fixed at 1066.4 nm. The measured optical parametric spectra are shown in Fig. 3.8. When the average pump power reaches 14.7 dBm (corresponding to a peak pump power of 55 W), the optical parametric sidebands can be clearly observed. With the increase of the pump power, the bandwidths of the Stokes and anti-Stokes sidebands are broadened. The frequency detuning between the pump and the signal or idler becomes large, and the

Fig. 3.7 The autocorrelation waveforms for the original and amplified pump pulses

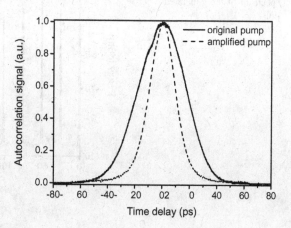

Fig. 3.8 The measured optical parametric gain sidebands versus the average pump power for the pump wavelength of 1066.4 nm

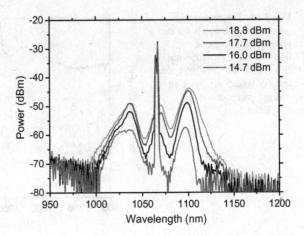

Fig. 3.9 The peak intensity of the Stokes sideband and the wavelength detuning between the Stokes sideband and the pump as a function of the average pump power

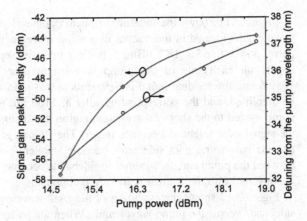

intensities of the signal and idler are increased. The extracted information about the frequency detuning and the peak intensity are shown in Fig. 3.9. When the average pump power increases from 14.7 to 18.8 dBm, the peak intensity of the Stokes sideband increases from −57.5 to −43.8 dBm, and the wavelength detuning between the peak of the Stokes sideband and the pump increases from 32.5 to 37.1 nm [12].

Then, the average pump power is fixed at 18.8 dBm (corresponding to a peak pump power of 140 W), and the pump wavelength is tuned from anomalous dispersion wavelength regime to near the ZDW. The output parametric spectra are shown in Fig. 3.10. The lobes on the anti-Stokes gain bands are attributed to the dispersive wave generation from the soliton generated by modulation instability. When the pump wavelength is far from the ZDW, the bandwidths of the Stokes and anti-Stokes are relatively narrow. After the pump moved close to the ZDW, both the Stokes and anti-Stokes parametric gain bands become broader.

Fig. 3.10 The measured parametric gain bands for the pump wavelengths of (i) 1066.4 nm, (ii) 1066 nm, (iii) 1065.6 nm, (iv) 1064 nm, and (v) 1062.5 nm, with an average pump power of 18.8 dBm

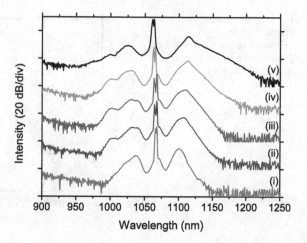

Figure 3.11 shows the measured optical parametric spectra for different pump wavelengths located in the normal dispersion wavelength regime with the average pump power fixed at 18.8 dBm. It is clear that two separated narrow gain bands appear on each side of the pump wavelength. For the pump wavelength of 1059.6 nm, the residual pump components at 974 nm from the YDF amplifier can be amplified, and the corresponding idler at 1160 nm is also amplified. With the pump moved to the shorter wavelength region, the detuning between the pump and the signal/idler sideband becomes larger. The Raman gain peak begins to emerge, and the parametric gain sidebands become smaller. This is because the walk-off between the pump and the parametric sidebands becomes larger. More pump power is converted into the Raman scattering components.

Figure 3.12 shows the evolution of the peak wavelength of the parametric gain sidebands versus the pump wavelength. When the pump wavelength is tuned from

Fig. 3.11 The measured parametric gain bands for the pump wavelengths of (i) 1061.6 nm, (ii) 1059.6 nm, (iii) 1057.5 nm, (iv) 1056.4 nm, and (v) 1055.6 nm, with an average pump power of 18.8 dBm

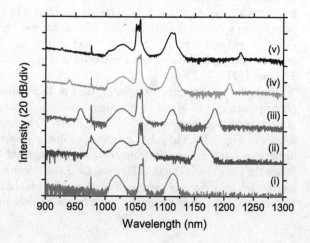

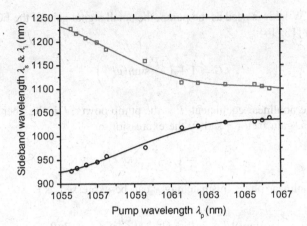

Fig. 3.12 The peaks of the parametric gain sidebands as a function of the pump wavelength around the ZDW of the PCF. The symbols are extracted from the experimental results, and the *solid line* represents the fitting curve

1055.6 to 1066.4 nm, the Stokes sideband can be tuned from 1228 to 1099 nm, and the anti-Stokes sideband can be tuned from 927 to 1038 nm. When the pump wavelength is shorter than 1055.6 nm, the FWM parametric sidebands cannot be observed. This is due to the limited peak pump power and the large walk-off between the pump and the signal/idler.

3.3 OPA Based on PCF Pumped by Picosecond Light Source

3.3.1 The Theory of Pulsed OPA

In the CW-pumped FOPA systems, in order to obtain a high gain, usually a long nonlinear gain fiber is needed. The FWM effect is sensitive to the ZDW of the fiber, and the parametric gain will be disturbed by the ZDW fluctuation along the fiber length. This requires a strict uniformity along the fiber length, which is relatively easy to be achieved in HNLF and DSF. In the case of PCF, complex index distributions are usually included in the cladding structures, and the ZDW drifting can be induced by slight disruptive factors. A high peak power can be easily obtained in the pulsed lasers, which are suitable to test the properties of the PCF. When a pulsed laser is used as the pump, the required fiber length can be largely reduced. Obvious signal amplification can be observed in a fiber with a length of dozens of meters, several meters, and even dozens of centimeters. In the pulse-pumped FOPA, the pump plays a role in a periodical time sections. In the time interval between the two periodical time sections, the pump has no effect to the amplification. It is necessary to revise the parametric gain calculating method.

When a CW laser is used as the pump, the small signal gain of the FOPA can be expressed as [13, 14]:

$$G = 1 + \left(\frac{\gamma P}{g}\sinh(gL)\right)^2 \tag{3.3}$$

where γ is the nonlinear coefficient, P is the pump power, L is the fiber length, and the parametric gain factor g satisfy the expression of

$$g^2 = (\gamma P)^2 - (\kappa/2)^2 \tag{3.4}$$

where the nonlinear phase mismatching κ can be expressed as

$$\kappa = \Delta\beta + 2\gamma P \approx \Omega^2\beta_2 + \frac{1}{12}\Omega^4\beta_4 + 2\gamma P \tag{3.5}$$

where Ω denotes the frequency detuning between the pump and the signal.

When a pulsed laser is used as the pump source, the pulse width of the pump pulse is supposed to be W, and the repetition rate is supposed to be R. The duty cycle of the pump pulse can be calculated to be

$$\eta_p = R \times W \tag{3.6}$$

A weak CW signal can be amplified to be pulses, the repetition rate of the amplified signal is the same with that of the pump, and the pulse width of the amplified signal is narrower than that of the pump pulse due to the pulse compression effect of the FOPA. Usually, the compression ratio is close to 0.2. Here, we suppose the compression ratio is C. The duty cycle of the amplified signal is

$$\eta_s = C \times R \times W \tag{3.7}$$

If the power of the CW signal is P_s, only a portion of the power $P_s \times \eta_s$ participate in the amplification process. The other portion of the signal power $P_s \times (1 - \eta_s)$ is just transmitted through the fiber. The amplified signal power is measured to be P_{as} at the output. The unamplified signal power should be removed, and the amplified signal power can be revised as

$$P_{ras} = P_{as} - P_s \times (1 - \eta_s) \tag{3.8}$$

The parametric gain can be calculated by

$$G = 10\log\left(\frac{P_{ras}}{P_s \times \eta_s}\right) \tag{3.9}$$

In the pulse-pumped FOPA, the required fiber length is short, so the fiber loss is neglected in the derivations [15].

3.3.2 High-Gain Broadband FOPA

In order to experimentally investigate the properties of the pulse-pumped FOPA, a
PCF with the ZDW located at 1064 nm is used as the nonlinear medium.
A picosecond mode-locked YDF laser is used as the pump source. The SEM image
of the cross section of the PCF and the dispersion curve for the fundamental mode
are shown in Fig. 3.13.

The experimental schematic for the PCF-based OPA is shown in Fig. 3.14. The
pump wavelength can be tuned from 1050 to 1070 nm, the repetition rate is

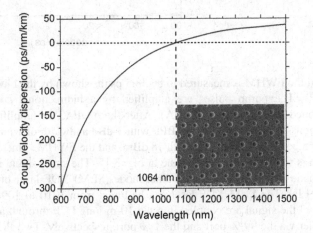

Fig. 3.13 The calculated group velocity dispersion for the fundamental mode of the PCF. The
inset shows the SEM image of the cross section of the PCF

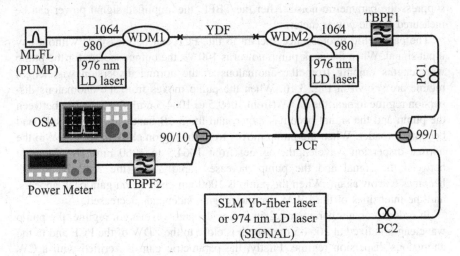

Fig. 3.14 The experimental schematic of the PCF-based optical parametric amplifier

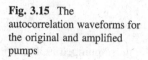

Fig. 3.15 The autocorrelation waveforms for the original and amplified pumps

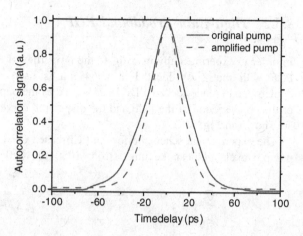

16 MHz, and the FWHM is measured to be 38.5 ps (as shown by the blue solid line in Fig. 3.15). The pump pulses are amplified by a bidirectional core-pumped ytterbium-doped fiber amplifier (YDFA). After the YDFA, the amplified spontaneous emissions are suppressed by a TBPF with 3-dB bandwidth of 1 nm. After the filter, the average pump power can reach 18 dBm, and the FWHM is measured to be 30 ps, which is shown by the red dot line in Fig. 3.15. The small signal is provided by a homemade tunable single-longitudinal-mode (SLM) YDF laser or a 974-nm laser diode (LD). The SLM YDF laser can be tuned from 1010 to 1090 nm [16]. The pump and the signal are coupled into the 10-m-long PCF through a 99/1 fiber optical coupler via the 99 % port and the 1 % port, respectively. Two PCs PC1 and PC2 are used to align the polarization states of the pump and the signal with respect to the principle axis of the PCF. At the output port, 10 % of the power is coupled into the OSA. Another TBPF is set after the 90 % port of the fiber coupler to suppress the parametric noise. After the TBPF, the amplified signal power can be measured by the power meter.

The parametric fluorescence spectra in this PCF can be measured without any input signal. When the peak pump power is 100 W, the output spectra for the pump wavelengths ranging from the anomalous to the normal dispersion wavelength regime are shown in Fig. 3.16. When the pump moves from the anomalous dispersion regime to near the ZDW (from 1069.5 to 1064.5 nm), the detuning between the pump and the signal becomes larger, and the 3-dB bandwidth of the sideband becomes broader. When the pump moves from the region close to the ZDW to the normal dispersion wavelength regime (from 1064.5 to 1060 nm), the detuning between the signal and the pump increases rapidly, and the 3-dB bandwidth becomes narrow again. When the pump is 1060 nm, the Raman gain peak appears, and the intensities of the FWM parametric gain sidebands decreased.

In order to obtain parametric gain in a wideband wavelength regime, the pump wavelength is fixed at 1064.5 nm which is close to the ZDW of the PCF and in the anomalous dispersion regime. Firstly, the parametric gain is verified with a CW

Fig. 3.16 The measured parametric ASE spectra for the pump wavelengths ranging from the anomalous to the normal dispersion wavelength regime

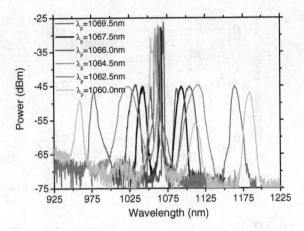

Fig. 3.17 The output spectrum of the FOPA pumped close to the ZDW of the PCF. The pump and the signal are operated at 1064.5 and 1032 nm, respectively

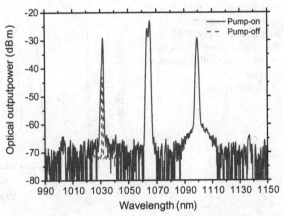

signal. Figure 3.17 shows the optical spectrum of the FOPA. The signal is set at 1032 nm. The signal is amplified significantly once the pump is launched into the PCF, and the corresponding idler is generated at 1099 nm. The cascaded FWM sidebands can also be observed at 1002 and 1136 nm, respectively. The average pump power is set to be 13.3 dBm (corresponding to a peak pump power of 45 W). When the pump is switched off, the signal power is measured to be −15 dBm at the output. When the pump is switched on, the signal power is measured to be −5.9 dBm. The pulse widths of the amplified signal and the generated idler are measured to be 8 and 6 ps, respectively, as shown in Fig. 3.18. Due to the exponential amplification effect of the FOPA, the pulse widths of the signal and idler are narrower than that of the pump. In the FWM process, the spectral width of the idler is usually wider than that of the signal because the pump has a certain spectral width. Hence, the idler has a narrower pulse width than the signal. The pulse compression ratios of the amplified signal and idler pulses are calculated to be 15/4 and 5/1, respectively. The duty cycle of the signal pulse can be calculated to be

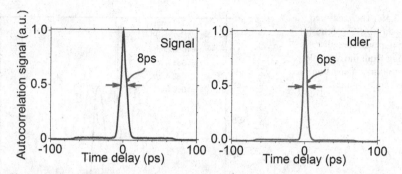

Fig. 3.18 The autocorrelation waveforms of the amplified signal and the generated idler

Fig. 3.19 The parametric
gain spectrum of the FOPA
for the pump at 1064.5 nm
with a peak power of 45 W.
The *filled squares* shows the
experimental results, and the
solid line shows the
simulation result

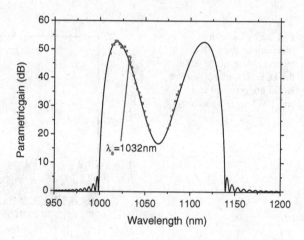

1/7812.5. The parametric gain at 1032 nm is calculated to be 47.4 dB by using
Eq. (3.9) as shown by the blue arrow in Fig. 3.19.

Then, the parametric gain bandwidth for the pump at 1064.5 nm is measured by
tuning the signal wavelength from 1010 to 1090 nm. The measured parametric gain
for different signal wavelengths is shown by the red filled squares in Fig. 3.19.
When the signal is located at 1017 nm, a peak gain of 52.9 dB is obtained. To
certificate the parametric gain, we theoretically calculate the gain profile by using
Eq. (3.3). For the pump wavelength of 1064.5 nm, $\beta_2 = -2.09 \times 10^{-4}$ ps^2/m and
$\beta_4 = -2.11 \times 10^{-9}$ ps^4/m. The effective length of the PCF is set to be 10 m by
neglecting the fiber loss. The simulation result is shown by the black solid line in
Fig. 3.19. It can be seen that the simulated and measured results agree well with
each other. In simulation, the parametric gain can be obtained in the wavelength
range from 999 to 1139 nm. The parametric gain bandwidth is roughly identical to
the fluorescence spectrum illustrated in Fig. 3.16. The evolution of the parametric
gain versus the peak pump power for the signal at 1017 nm is shown in Fig. 3.20.
When the pump power is relatively low, the parametric gain grows monotonously

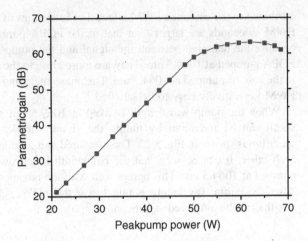

Fig. 3.20 The parametric gain with respect to the peak pump power for the signal at 1017 nm

with the peak pump power. After the parametric gain reaches a maximal value of 62.2 dB, it begins to decrease with the peak pump power. This is because the FOPA is saturated by the dramatic increase of the parametric fluorescence.

3.3.3 Widely Tunable Parametric Amplification

According to the fluorescence spectra shown in Fig. 3.16, the parametric gain can be tuned by adjusting the pump wavelength. Then, the FOPA is operated with the pump located at 1069.5 nm and the signal located at 1057.5 nm. The output spectrum of the FOPA is shown in Fig. 3.21. It can be seen that the signal is significantly amplified, and the corresponding idler is efficiently generated at 1082 nm. The optical components at 1045.5, 1095, 1034, and 1107.5 nm are

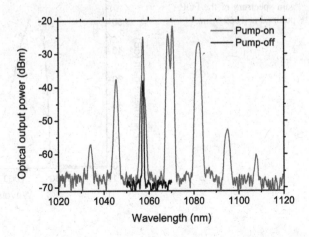

Fig. 3.21 The optical spectrum of the FOPA pumped at 1069.5 nm in the anomalous dispersion wavelength regime of the PCF

attributed to the cascaded FWM effect. The intensities of the second-order cascaded FWM sidebands are larger than that in the FOPA pumped at 1064.5 nm. This is because the detuning between the signal and the pump is smaller than that of the FOPA pumped at 1064.5 nm. They are more close to the ZDW of the PCF than that in the case of pumped at 1064.5 nm. The phase-matching condition for the cascaded FWM is relatively easy to be satisfied.

When the pump wavelength is fixed at 1069.5 nm, the parametric gain bandwidth can be measured by tuning the signal wavelength. The parametric gain spectrum is shown in Fig. 3.22. The measured and calculated results agree well with each other. It can be seen that the bandwidth is narrower than that of the FOPA pumped at 1064.5 nm. This agrees well with the parametric fluorescence spectrum. After the pump wavelength is tuned near the ZDW, signals with different wavelengths can be amplified significantly.

3.3.4 FOPA-Based Picosecond Pulse Generation

From the parametric fluorescence spectra, we can see that when the pump wavelength is tuned, the parametric gain can be formed in new wavelength regions. The parametric gain can cover a wide wavelength band by tuning the pump wavelength. When a CW signal locates in the parametric gain band, the CW signal can be modulating amplified to be pulses by the pump wave, and the corresponding idler pulses can be generated simultaneously. Figure 3.23 shows the optical spectrum of the FOPA pumped at 1062.5 nm. The signal at 974 nm is significantly amplified, and the idler is generated at 1168 nm. The pulse widths of the signal and idler pulses are measured to be 9 and 6.5 ps, respectively, which are shown in Fig. 3.24.

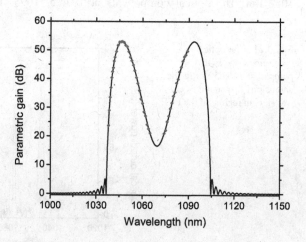

Fig. 3.22 The parametric gain spectrum of the FOPA pumped at 1069.5 nm

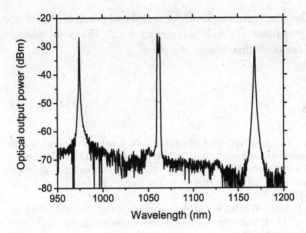

Fig. 3.23 The optical spectrum of the FOPA pumped at 1062.5 nm in the normal dispersion wavelength regime

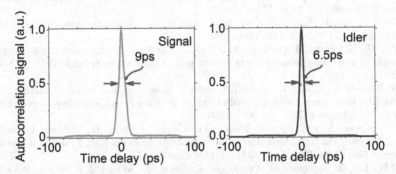

Fig. 3.24 The autocorrelation waveforms of the amplified signal and the corresponding idler

3.4   Conclusion

In this chapter, in order to use the YDF laser as the pump source, the dispersion-flattened PCF with the ZDW located near 1064 nm is designed and fabricated. The FWM parametric sidebands are observed in the PCF. The central wavelength of the Stokes sideband can be tuned from 1099 to 1228 nm, and the central wavelength of the anti-Stokes sideband can be tuned from 927 to 1038 nm.

An FOPA is built up by coupling the picosecond pump and CW signal into the PCF simultaneously. The method to calculate the parametric gain for the pulse-pumped FOPA is developed. When the pump locates near the ZDW, high gain can be obtained in a wide band. After the pump wavelength is tuned near the ZDW, the parametric gain can cover a wide wavelength region.

New wavelength pulses can be generated in the PCF-based OPA. The CW signal can be amplified to picosecond pulse, and the picosecond idler can also be generated. When the wavelengths of the signal and pump are tuned, the picosecond pulse can be generated in a wide wavelength band. The pulse widths of the signal and idler are narrower than that of the pump.

References

1. M.E. Marhic, K.K.Y. Wong, G. Kalogerakis, L.G. Kazovsky, Toward practical fiber optical parametric amplifiers and oscillators. Opt. Photonics News **15**, 20–25 (2004)
2. M.E. Marhic, Y. Part, F.S. Yang, L.G. Kazovsky, Broadband fiber-optical parametric amplifiers and wavelength converters with low-ripple Chebyshev gain spectra. Opt. Lett. **21** (17), 1354–1356 (1996)
3. J. Hansryd, P.A. Andrekson, Broad-band continuous-wave-pumped fiber optical parametric amplifier with 49-dB gain and wavelength-conversion efficiency. IEEE Photonics Technol. Lett. **13**(3), 194–196 (2001)
4. T. Torounidis, P.A. Andrekson, B.E. Olsson, Fiber-optical parametric amplifier with 70-dB gain. IEEE Photonics Technol. Let. **18**(10), 1194–1196 (2006)
5. M.E. Marhic, N. Kagi, T.K. Chiang, L.G. Kazovsky, Broadband fiber optical parametric amplifiers. Opt. Lett. **21**(8), 573–575 (1996)
6. T. Torounidis, P. Andrekson, Broadband single-pumped fiber-optic parametric amplifiers. IEEE Photonics Technol. Lett. **19**(9), 650–652 (2007)
7. M.C. Ho, K. Uesaka, M. Marhic, Y. Akasaka, L.G. Kazovsky, 200-nm-bandwidth fiber optical amplifier combining parametric and Raman gain. J. Lightwave Technol. **19**(7), 977–981 (2001)
8. W. Wadsworth, N. Joly, J. Knight, T.A. Birks, F. Biancalana, P.J. Russell, Supercontinuum and four-wave mixing with Q-switched pulses in endlessly single-mode photonic crystal fibres. Opt. Express **12**(2), 299–309 (2004)
9. T. Sylvestre, A. Kudlinski, A. Mussot, J.F. Gleyze, A. Jolly, H. Maillotte, Parametric amplification and wavelength conversion in the 1040–1090 nm band by use of a photonic crystal fiber. Appl. Phys. Lett. **94**(11), 111104 (2009)
10. M.W. Lee, T. Sylvestre, M. Delqué, A. Kudlinski, A. Mossot, J.-F. Gleyze, A. Jolly, H. Maillotte, Demonstration of an all-fiber broadband optical parametric amplifier at 1 μm. J. Lightwave Technol. **28**(15), 2173–2178 (2010)
11. A. Mussot, A. Kudlinski, R. Habert, I. Dahman, G. Mélin, L. Galkovsky, A. Fleureau, S. Lempereur, L. Lago, D. Bigourd, T. Sylvestre, M.W. Lee, E. Hugonnot, 20 THz-bandwidth continuous-wave fiber optical parametric amplifier operating at 1 μm using a dispersion-stabilized photonic crystal fiber. Opt. Express **20**(27), 28906–28911 (2012)
12. L. Zhang, S.-G. Yang, X.-J. Wang, D.-D. Gou, H.-W. Chen, M.-H. Chen, S.-Z. Xie, Demonstration of optical parametric gain generation at 1 μm regime based on a photonic crystal fiber pumped by picosecond mode-locked ytterbium-doped fiber laser. J. Opt. **16**(1), 015202 (2014)
13. R.H. Stolen, J.E. Bjorkholm, Parametric amplification and frequency conversion in optical fibers. IEEE J. Quantum Electron. **18**(7), 1062–1072 (1982)
14. T. Torounidis, M. Karlsson, P.A. Andrekson, Fiber optical parametric amplifier pulse source: Theory and experiments. J. Lightwave Technol. **23**(12), 4067–4073 (2005)
15. L. Zhang, S. Yang, X. Wang, D. Gou, W. Chen, W. Luo, H. Chen, M. Chen, S. Xie, Photonic crystal fiber based wavelength-tunable optical parametric amplifier and picosecond pulse generation. IEEE Photonics J. **6**(5), 1501908 (2014)
16. F. Yin, S. Yang, H. Chen, M. Chen, S. Xie, 60-nm-wide tunable single-longitudinal-mode ytterbium fiber laser with passive multiple-ring cavity. IEEE Photonics Technol. Lett. **23**(22), 1658–1660 (2011)

Chapter 4
Widely Tunable Optical Parametric Oscillator (OPO) Based on PCF

4.1 Introduction

Not long after the laser was invented, optical nonlinearity was used to generate high-intensity radiation at new wavelength. In 1965, Wang and Racette discovered that new wavelength laser can be generated by two beams of light transmitted in a nonlinear crystal. The wavelength of the newly generated laser was just the frequency difference of the two pump lights [1]. In the same year, Giordmaine and Miller demonstrated an optical parametric oscillator (OPO) in a LiNbO$_3$ crystal pumped by a pulsed laser [2]. Afterward, a lot of researchers focused on the investigation of OPO. When the phase-matching condition is satisfied, parametric process is recognized as the primary nonlinear effect to occur [3]. OPO has been recognized as an important technique to generate tunable coherent light at novel wavelength regions because of their properties of widely tunable range, high efficiency, and ease of operation. The $\chi^{(2)}$ nonlinear effect of many crystals has been utilized to develop OPO. However, the crystal OPO requires a precise alignment within the cavity for the parametric oscillation. It is very inconvenient for the implement.

Recently, the fiber optical parametric oscillator (FOPO), which utilizes the parametric gain generated in the nonlinear optical fiber, has attracted comprehensive interests. The fibers in the cavity can be easily spliced, and the precise alignment can be omitted. Furthermore, the fiber systems can be operated steadily in varieties of harsh conditions. Serkland et al. reported a FOPO in a nonlinear fiber Sagnac interferometer pumped by a 1544-nm pulsed laser. The pulse widths of the pump and the output of the oscillator are 3.9 and 0.83 ps, respectively. The pulse compression effect is discovered experimentally in the oscillator [4]. Marhic et al. [5] demonstrated a FOPO in a 100-m-long HNLF pumped by a CW laser with a power of 240 mW. The output wavelength can be tuned from 1500 to 1580 nm. Zhou et al. [6] achieved a widely tunable FOPO in a 50-m-long HNLF pumped by a picosecond erbium-doped fiber laser. The output wavelength can be tuned over

© Springer-Verlag Berlin Heidelberg 2016
L. Zhang, *Ultra-Broadly Tunable Light Sources Based on the Nonlinear Effects in Photonic Crystal Fibers*, Springer Theses,
DOI 10.1007/978-3-662-48360-2_4

250 nm by adjusting the pump wavelength. Yang et al. demonstrated an actively mode-locked FOPO by inserting an amplitude modulator (AM) into the oscillating cavity. The wavelength of the pulse train can be tuned over 21 nm by adjusting the intra-cavity filter [7]. In the C and L communication bands, using the state-of-the-art DSF or HNLF as the gain medium, the CW and pulsed FOPOs can be well developed.

Thanks to the appearance of the PCF, the ZDW can be moved to a shorter wavelength region. Benefiting from the high nonlinearity of the PCF, a FOPO with better performance can be achieved potentially. In 2002, Sharping et al. reported an OPO in a PCF pumped by a Ti: sapphire femtosecond laser. After the peak pump power reaches 34.4 W, the oscillation can be built up. The output wavelength can be tuned from 725 to 780 nm [8]. In 2004, de Matos et al. [9] reported an OPO with a wavelength tuning range of 50 nm in a PCF pumped by a CW laser at 1550 nm. In 2005, Deng et al. demonstrated an OPO in a 65-cm-long PCF pumped by a mode-locked ytterbium-doped fiber laser. The output pulse width is compressed to 460 fs, and the wavelength can be tuned over 200 nm [10]. In 2008, Sharping et al. [11] reported another PCF-based OPO with a high output power of 50 mW.

Most of these PCF-based OPOs include a lot of free-space optical components (lens and mirrors). The stability and operability are heavily discounted. With the improving of the fiber drawing and splicing technologies, it is possible to build the PCF-based OPO in all fiber structures. In this chapter, we focus to investigate the all-fiber-integrated OPO pumped near 1064 nm.

4.2   Wavelength Tuning Methods

One of the advantages of the OPO is that highly coherent radiation can be obtained at novel wavelength region. In the oscillating cavity of the FOPO, the gain is provided by the FWM in the optical parametric process. The parametric components are fed back into the gain fiber by the cavity structure. With the gain larger than the cavity loss, the oscillation can be set up. The output wavelength of the FOPO can be tuned by adjusting the wavelength inside parametric gain band. Furthermore, the FWM parametric gain band generated in the optical fiber has certain amount of bandwidth. The output wavelength of the FOPO can also be tuned by designing special cavity structures to make different spectral components in the FWM gain band can be selected to oscillate.

In the first kind of wavelength tuning scheme, the key point is to adjust the wavelength of the FWM gain band. According to the phase-matching condition, the wavelength of the gain band is related to the even-order dispersion, nonlinear coefficient, and the pump power. Here, we list five kinds of method to tune these parameters at the pump wavelength. The first method is to tune the pump wavelength. In Ref. [12], the FWM can be tuned in a wide wavelength range by tuning the pump wavelength. The second method is to tune the polarization state of the pump wave, which is specially suitable for the PCF-based OPO. The parameters of

the PCF can be largely tuned by adjusting the polarization state. In Ref. [13], the
FWM gain band is well tuned by the polarization state. The third method is to
adjust the pressure, tension, or temperature of the gain fiber. The influence of the
pressure, tension, and temperature to the ZDW of the SMF has been demonstrated
as early as 1992 [14]. In 1998, Ghosh et al. [15] reported the Sellmeier coefficients
of the fiber material can be changed by the pressure. It is an effective approach to
tune the parameters of the PCF. The fourth method is to adjust the value of the
nonlinear phase mismatch term by tuning the pump power. Velazquez-Ibarra et al.
[16] ever demonstrated the influence of the pump power to the FWM wavelength.
The fifth method is to taper a fiber. The dispersion and nonlinear coefficient can be
continuously changed simultaneously.

In the second kind of wavelength tuning scheme, the key point is to design the
cavity structures to make different wavelength components oscillated. To the cavity
structures, different people have different views. Here, we list two kinds of cavity
structures. The first kind of the cavity structures is to insert an optical filter component (tunable band-pass filter or fiber Brag grating) in the oscillating cavity. Lasri
et al. [17] ever used this kind of structure to set up a high-speed pulsed FOPO. The
second kind of the cavity structures is to use the time-dispersion-tuned technique.
Different FWM parametric components can be tuned to oscillate by adjusting the
cavity length. It can be only used in the pulsed pump case, which was firstly used in
the Raman fiber laser [18]. Recently, it is applied to the FOPO systems [19, 20].
The principle of the time-dispersion-tuned technique is to stretch the pulse of the
FWM parametric gain band in time domain by the group velocity dispersion of the
oscillating cavity. Different wavelength components in the parametric gain band
can be tuned to be synchronized with the succedent pump pulses after a round-trip
in the cavity by slightly adjusting the cavity length. During the time-dispersion-
tuning process, the round-trip time of different wavelength components located in
the FWM parametric gain band is different. The time difference can be expressed as

$$\Delta T = |t(\lambda_1) - t(\lambda_2)| = \left| \sum_i L_i D_i(\lambda_c) |\lambda_1 - \lambda_2| \right|, \quad i = p, s, a \qquad (4.1)$$

where $t(\lambda)$ is the round-trip time of the optical signal at the wavelength of λ, λ_1, and
λ_2 in the FWM parametric gain band, and $\lambda_c = (\lambda_1 + \lambda_2)/2$. L_p, L_s, and L_a are the
length of PCF, single-mode fiber (SMF), and air in the oscillating cavity, respec-
tively. D_p, D_s, and D_a (λ_c) are the group velocity dispersion of the PCF, SMF, and
air, respectively. The time difference ΔT of two signal components with slightly
different wavelengths can be enlarged, and the wavelength resolution of the output
of the FOPO can be enhanced by a large cavity dispersion. The oscillation of the
FOPO can be achieved by coarse tuning the cavity length, and the output wavelength of the FOPO can be continuously tuned by fine-tuning the cavity length.

4.3 Wavelength Tunability via Tuning of the Pump Wavelength

FOPO is an effective method to generate tunable laser at new wavelength. The ZDWs of the traditional HNLFs and DSFs are located above 1310 nm. In the C and L telecommunication bands, using the DSF and HNLF as the gain medium, CW and pulsed FOPO have been reported comprehensively. Thanks to the appearance of the PCF, the ZDW can be moved to shorter wavelength region. FWM sidebands have been demonstrated by pumping at visible and 1 μm regions, respectively. FOPOs based on the PCFs pumped at visible and 1 μm regions have also been demonstrated. However, most of the reports are focused on the situations pumped by femtosecond pulses. It has very short interaction length due to the walk-off between the pump and signal or idler. This requires the gain PCF is very short and free-space optical components are used. It is difficult to integrate with other fiber components. Recently, a tunable CW all-fiber OPO pumped at 1 μm region is demonstrated. The CW pump power is limited at 11 W, and the output of the FOPO covers the range from 950 nm to 1010 nm. Picosecond pulse has wider pulse width than femtosecond pulse, so longer gain fiber can be used. Picosecond pulse with a peak power of hundreds watt is easy to be obtained at 1 μm region. It is possible to extend the tuning range of the FOPO significantly. Thus, a FOPO based on a PCF pumped by a picosecond pulse train at 1 μm region is investigated in this section.

The experimental setup is shown in Fig. 4.1. The pump source is a homemade tunable mode-locked ytterbium-doped fiber laser. Its wavelength can be tuned from

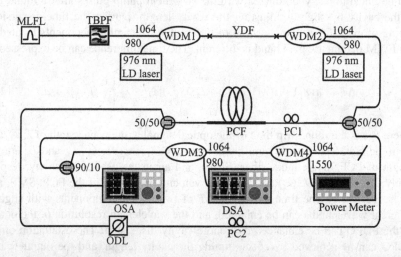

Fig. 4.1 The experimental setup of the PCF-based OPO. *MLFL* mode-locked ytterbium-doped fiber laser; *TBPF* tunable band-pass filter; *WDM* wavelength division multiplex; *YDF* ytterbium-doped fiber; *PC* polarization controller; *PCF* photonic crystal fiber; *ODL* optical delay line; *OSA* optical spectrum analyzer; *DSA* digital serial analyzer. Reprinted from Ref. [22], copyright © OSA 2013, with permission from OSA

1033 to 1070 nm, the repetition rate is 25.4 MHz, and the full width at half maximum (FWHM) is 35 ps. After filtered by a tunable band-pass filter (TBPF) with 3-dB bandwidth of 1 nm, the original pulse is amplified by a bidirectional core-pumped fiber amplifier using 1-m ytterbium-doped fiber (YDF). The FWHM of the amplified pulse is compressed to 21 ps. The compression of the pulse can be understood by noting that the pulse spectrum can be expanded as long as the gain in the spectral wings exceeds the loss level. Then, the pulse train is coupled into the cavity through a 50/50 coupler. Compared with the specialty dichroic mirror or WDM coupler used in previous reports, in our scheme, all the spectral components in the gain band can be fed back into the cavity and the cost of the system can be largely reduced. The consequent advantage is that the oscillating wavelength can cover the whole gain bandwidth of the OPA gain spectrum. A polarization controller PC1 is employed to align the polarization state of the pump wave with the principle axis of the PCF. The cavity length is about 18 m, including a 10-m-long PCF as the gain medium, with a nonlinear coefficient of 15 $W^{-1} km^{-1}$ and the ZDW located at 1062 nm. The signal and idler inside the cavity can be closely synchronized with the succedent pump pulse after a round-trip in the 18-m cavity. The splice loss between the Hi-1060 SMF and the PCF is optimized by inserting a piece of ultra-high numerical aperture (UHNA) fiber between them, and the mode fields are well matched. After the PCF, a 50/50 coupler provides 50 % feedback and 50 % output. In the feedback branch, an optical delay line (ODL) is used to adjust the cavity length. Both the signal (longer than pump) and idler (shorter than pump) are launched back into the cavity. Either the signal or the idler can be selected to oscillate by fine-tuning the ODL. The polarization controller PC2 is used to align the polarization state of the light wave inside the cavity. The FOPO output spectrum is measured by an optical spectrum analyzer (OSA) through a 10/90 coupler. At the output port of the FOPO, the idler pulses below 1 μm can be filtered out through the WDM3, and the signal pulses above 1180 nm can be filtered out through the WDM4. The waveforms of the signal (idler) can be recorded by a digital serial analyzer (DSA), and the powers of the signal (idler) can be measured by a power meter.

In the FOPO configuration, the gain is provided by the FWM process in the PCF, which is usually pumped near the ZDW. The appearance of the FWM requires the phase-matching condition to be satisfied. The wavelength of the parametric gain band can be tuned by adjusting the pump wavelength near the ZDW of the PCF. The gain is measured to be about 30 dB with the pump wavelength of 1063 nm and a small signal at the wavelength of 1030 nm. The round-trip loss of the cavity is measured to be 8 dB at the wavelength of 1035 nm. The gain is much larger than the round-trip loss in the cavity, so the operation of the FOPO at a wide wavelength range is possible.

The experimentally measured optical spectra of the output of the oscillator for several different pump wavelengths near the ZDW are depicted in Fig. 4.2. When the pump wavelength moves from 1066.3 to 1057.5 nm, the signal pulse can be tuned from 1086.5 to 1195.8 nm, and the idler pulse can be tuned from 1047 to 949 nm. The lobe at the wavelength region from 1015 to 1045 nm is introduced by

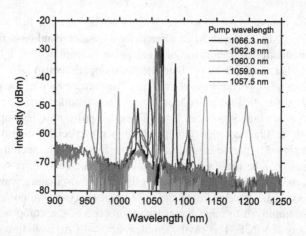

Fig. 4.2 Optical spectra of the output of the oscillator for the pump wavelengths of 1066.3, 1062.8, 1060.0, 1059.0, and 1057.5 nm. For the pump wavelength of 1060.0 nm, the idler wave is chosen to synchronize with the pump. For the other pump wavelengths, the signal wave is chosen to synchronize with the pump. Reprinted from Ref. [22], copyright © OSA 2013, with permission from OSA

the bidirectional core-pumped ytterbium-doped fiber amplifier. For the pump wavelength of 1066.3 nm, the high-order FWM peaks are also observed. When the pump wavelength is varied, the ODL should be tuned to synchronize the signal or the idler with the pump by adjusting the cavity length. For the pump wavelength of 1060.0 nm, the idler wave is chosen to be synchronized with the pump, and the intensity peak of the idler is higher than that of the signal. For the other pump wavelength, the signal wave is selected to synchronize with the pump, so the intensity peak of the signal is higher than that of the idler. The frequency detuning of the signal or idler with respect to the pump wave increases as the pump wavelength decreases. For the pump wavelength of 1056 and 1053 nm, the optical spectra of the output of the oscillator are shown in Fig. 4.3. The wavelength of the signal pulse can be tuned up to 1277 nm, and the wavelength of the idler pulse can be tuned down to 898 nm. Except for the signal and idler peaks, the Raman gain peak is also observed. It is especially surprising that the Raman gain peak is larger than the gain peak of the signal or idler. The maximum Raman gain is generally believed to be smaller than the gain of phase-matched parametric process [21]. To some extent, it appears that the efficiency of the FWM is largely reduced by the increased walk-off effect and a large portion of the pump powers are converted to the Raman gain peak, when the pump wave is far from the ZDW in the normal dispersion regime [22].

Figure 4.4 shows the optical spectra of the output of the oscillator for the pump wavelength of 1057.8 nm with different pump power. It is clear that the bandwidth of the output of the FOPO is narrower with a small pump power and the bandwidth of the signal or idler becomes broader with the increasing of the pump power. The 3-dB bandwidth of the signal (idler) increases from 4.15 (3.9) to 16.7 (10.15) nm,

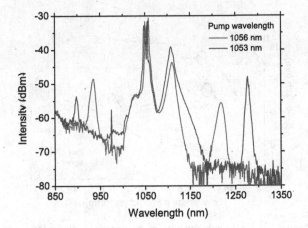

Fig. 4.3 Optical spectra of the output of the oscillator for the pump wavelengths of 1056 and 1053 nm. The peaks at 1110 nm are induced from the Raman gain. The peaks at 975 nm are from the semiconductor laser diode (LD laser) of 975 nm. Reprinted from Ref. [22], copyright © OSA 2013, with permission from OSA

Fig. 4.4 Optical spectra of the output of the oscillator for the pump wavelength of 1057.8 nm with different pump power. Reprinted from Ref. [22], copyright © OSA 2013, with permission from OSA

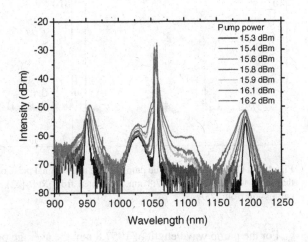

with the average pump power increasing from 15.3 to 16.2 dBm. The intensity peak of the signal or idler increases initially and then decreases with the increasing of the pump power. The bandwidth and the intensity peak as a function of the average pump power are depicted in Fig. 4.5. The idler pulse can be filtered out by the 980-nm port of the WDM3, and the signal pulse can be filtered out by the 1550 nm port of the WDM4. The waveforms of the input pump pulse at 1057.8 nm and the output signal pulse at 1193.5 nm measured from the DSA are shown in Fig. 4.6. It can be seen that the pulse width of the signal is narrower than that of the pump. This is induced by the pulse compression effect. The precise pulse width cannot be measured by the DSA. This is limited by the photodetector electrical bandwidth of 1.2 GHz.

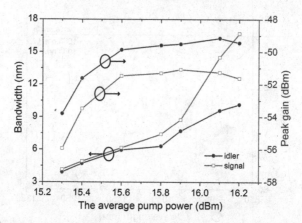

Fig. 4.5 The bandwidth and the intensity peak of the signal and idler pulses as a function of the average pump power. Reprinted from Ref. [22], copyright © OSA 2013, with permission from OSA

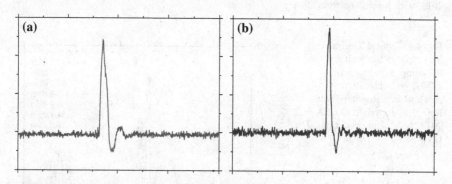

Fig. 4.6 **a** Output pulse from the pump source and **b** signal pulse observed at the FOPO output for the pump wavelength of 1057.8 nm. Reprinted from Ref. [22], copyright © OSA 2013, with permission from OSA

For the pump wavelength of 1057.8 nm, the average powers of the signal or idler from the output of the FOPO are measured by a power meter. The results are shown in Fig. 4.7. With the average pump power increased from 15.3 dBm to 16.2 dBm, the signal power increases from −6 to −1.55 dBm, the idler power increases from −4.8 to −2.9 dBm, and the corresponding conversion efficiency of the signal (idler) increases from 0.74 % (0.98 %) to 1.68 % (1.23 %). However, the power increasing rate was slowing down, as the average pump power is larger than 15.9 dBm, because of the gain saturation.

An all-fiber widely tunable FOPO based on a PCF pumped by a picosecond ytterbium-doped fiber laser has been demonstrated in this section. With the pump wavelength tuned between 1066.3 and 1053 nm, the output wavelength of the oscillator can be continuously tuned from 898 to 1047 nm and from 1086 to

Fig. 4.7 The average powers of the signal and idler from the output of the FOPO as a function of the average pump power for the pump wavelength of 1057.8 nm. Reprinted from Ref. [22], copyright © OSA 2013, with permission from OSA

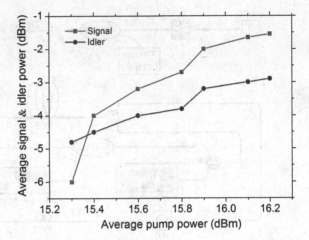

1277 nm. When the pump wavelength is far from the ZDW in the normal dispersion regime, the Raman gain peak is larger than the FWM gain peaks. The bandwidth of the output of the oscillator can be tuned by slightly varying the average pump power. This scheme is useful in generating picosecond pulse at nonconventional wavelength ranges around 1 μm.

4.4 Time-Dispersion-Tuned OPO Based on PCF

In the last section, the output of the PCF-based OPO can be tuned in a wavelength range of 340 nm by adjusting the pump wavelength in a range of 13.3 nm near the ZDW. It is clear that adjusting the pump wavelength is an effective way to tune the output wavelength of the FOPO. However, the oscillating wavelength of FOPO is usually too sensitive to the pump wavelength. For example, the output wavelength of the FOPO will be detuned by tens of nanometers (even hundreds of nanometers) away simply by adjusting the pump wavelength by about only one nanometer in the normal dispersion regime. Hence, it is very difficult to tune the pump wavelength rigorously so as to continuously tune the output wavelength of the FOPO. Furthermore, once the pump wavelength is changed, the optical components inside the FOPO cavity (such as the filter) should be adjusted accordingly. Hence, it is a hard work to tune the wavelength of the FOPO to cover a broad wavelength band by only adjusting the pump wavelength. In this section, we focus to optimize the wavelength tunability of the FOPO. Different wavelength components can be tuned to synchronize with the pump pulse by the time-dispersion-tuned technique. This cavity structure makes the output wavelength of the FOPO to be continuously tuned easily.

The experimental setup of the time-dispersion-tuned FOPO is shown in Fig. 4.8. The optical components used in the system are similar to that of the FOPO shown in

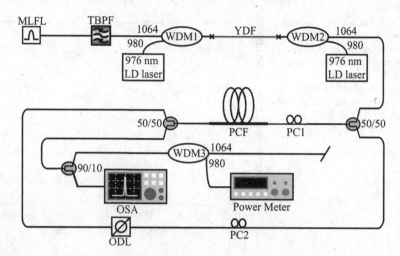

Fig. 4.8 Experimental setup of the time-dispersion-tuned FOPO. Reprinted from Ref. [20], copyright © OSA 2013, with permission from OSA

the last section. However, in order to utilize the time-dispersion-tuned technique, a long cavity length is used. In last section, in order to synchronize the FWM parametric components with the pump closely, a short cavity is used. In this section, the FWM parametric gain band needs to be broadened in the time domain. Large intra-cavity dispersion is required, so a long cavity of 64 m including a 50-m-long PCF is used in the system. The ZDW of the PCF is located at 1064 nm. The nonlinearity coefficient γ is calculated to be about 15 W^{-1} km^{-1}, and the fiber loss is measured to be 25 dB/km by the cutback method at the wavelength of 1064 nm.

First of all, in order to measure the FWM parametric gain band, the cavity was disconnected at a port of the ODL to form a fiber optical parametric generator. Figure 4.9 shows the parametric gain band for the pump wavelength of 1064.8 nm with average power of 11 dBm at the input port of the PCF. Two sidebands can be observed locating on the either side of the pump wave. Then, the FOPO is built up by connecting the cavity. The FWM sideband is tuned to synchronize with the succedent pump pulse after a round-trip in the cavity by coarse tuning the cavity length, and the FOPO begins to oscillate. The threshold average pump power injected into the PCF is measured to be 10 dBm. When the average pump power is increased to 11 dBm, the high-intensity narrow bandwidth output is obtained. The FWM signal sideband is located in the anomalous dispersion wavelength regime of the cavity. Different wavelength components propagate at different group velocities. The round-trip time difference between different wavelength components is enlarged by the cavity dispersion. The different wavelength components in the FWM sideband can be selected to synchronize with the pump pulse by fine-tuning the cavity length. The output wavelength of the FOPO can be continuously tuned. The measured spectra of the output of the FOPO with the pump wavelength fixed at 1064.8 nm are shown in Fig. 4.10. The signal pulse can be continuously tuned from

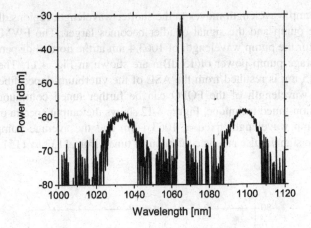

Fig. 4.9 The FWM parametric gain spectrum of the optical parametric generator for the pump wavelength of 1064.8 nm at the anomalous dispersion regime of the PCF. Reprinted from Ref. [20], copyright © OSA 2013, with permission from OSA

Fig. 4.10 The optical spectra of the continuously tuned output of the FOPO for the pump wavelength of 1064.8 nm. Reprinted from Ref. [20], copyright © OSA 2013, with permission from OSA

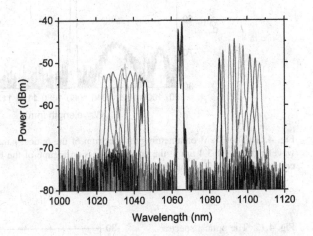

1085 nm to 1107 nm and the idler pulse can be continuously tuned from 1023 to 1046 nm by tuning the ODL in a range of about 12 mm, corresponding to 40 ps time delay. The wavelength tuning range is roughly identical to the parametric gain band. In the case of $\lambda_1 = 1085$ nm and $\lambda_2 = 1107$ nm, λ_c will be 1096 nm. Then, the time difference ΔT is calculated to be 36 ps, which is matched well with the experimental result of 40 ps. The wavelength tuning range is roughly identical to the FWM parametric sidebands. The peak of the signal is higher than that of the idler, while the linewidth of the signal is narrower than that of the idler, due to that the signal is the spectral component that has been synchronized with the pump pulses, not the idler.

As the pump wavelength moves to the shorter wavelength regions, the detuning between the pump and the signal or idler becomes larger. The FWM parametric gain bands for the pump wavelength of 1063.4 nm at the normal dispersion regime with an average pump power of 15 dBm are shown in Fig. 4.11. The lobe from 1015 to 1045 nm is resulted from the ASE of the ytterbium-doped fiber amplifier. The output wavelength of the FOPO can be further tuned continuously by the time-dispersion-tuned technique. Figure 4.12 shows the output spectra of the FOPO with the pump wavelength fixed at 1063.4 nm and the average pump power of 15 dBm. The signal pulse can be continuously tuned from 1133 to 1151 nm and the

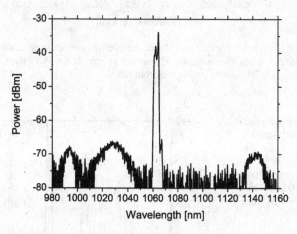

Fig. 4.11 The FWM parametric gain spectrum of the optical parametric generator for the pump wavelength of 1063.4 nm at the normal dispersion regime of the PCF. Reprinted from Ref. [20], copyright © OSA 2013, with permission from OSA

Fig. 4.12 The optical spectra of the continuously tunable output of the FOPO for the pump wavelength of 1063.4 nm. Reprinted from Ref. [20], copyright © OSA 2013, with permission from OSA

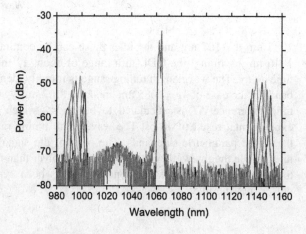

idler pulse can be continuously tuned from 988 to 1002 nm. The wavelength tuning range was also roughly identical to the parametric gain bands.

The output wavelength of the FOPO can be continuously tuned in a broadband by the combination of adjusting the pump wavelength and the time-dispersion-tuned technique. At first, the wavelength of 1151 nm is oscillated, and the oscillation can be moved to the shorter wavelength by increasing the optical delay time. The evolution of the output wavelength versus the relative optical delay time with respect to the oscillating wavelength of 1151 nm is shown in Fig. 4.13. With the relative optical delay time varied in a range of 100 ps, the output wavelength can be tuned from 988 to 1046 nm and from 1085 to 1151 nm, respectively. Except for the time dispersion tuning, the pump wavelength is also adjusted (1063.4, 1063.6, 1063.9, and 1064.8 nm), which are shown by the blue signs in Fig. 4.13.

At the output port of the FOPO, the idler pulses below 1 μm can be filtered out through the WDM3. The average power and bandwidth versus the idler wavelength is shown in Fig. 4.14. The average power fluctuates in the range from −5.5 to −4.5 dBm, which corresponding to the conversion efficiency (idler FOPO output power/input pump power) from 0.89 to 1.12 %, and the bandwidth drifts in the range from 1.0 to 3.72 nm, when the wavelength of the idler pulse is located in the wavelength range from 988 to 998 nm. The signal and idler bandwidths over the full tuning range drift in the ranges from 0.6 to 1.9 nm and from 1 to 3.72 nm, respectively, which are comparative with the bandwidth of the pump source. When the pump is fixed at one wavelength, the output wavelength of the FOPO can be tuned by adjusting the ODL. With the tuning of the ODL, an envelope can be formed by the intensity peak traces, which is matched with the parametric gain band profile. In the case of the oscillating wavelength located at the spectral edge of the parametric gain band, the average power of the output pulse is smaller, and the

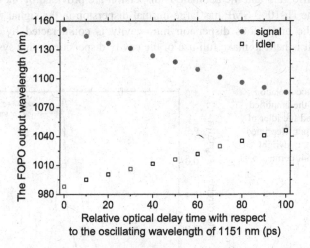

Fig. 4.13 Output wavelength of the FOPO versus the relative optical delay time with the oscillating wavelength of 1151 nm. The *blue signs* mean the adjusting of the pump wavelength. Reprinted from Ref. [20], copyright © OSA 2013, with permission from OSA

Fig. 4.14 The average idler power and 3-dB bandwidth as a function of the output idler wavelength. Reprinted from Ref. [20], copyright © OSA 2013, with permission from OSA

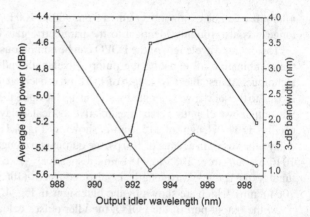

bandwidth is broader. Then, the properties of the output pulse can be improved by tuning the pump wavelength. Thus, a high-intensity peak and narrow linewidth output can be achieved in a broadband region except for the outermost edge by the combination of adjusting the pump wavelength and the time-dispersion-tuned technique. The autocorrelation trace of the idler below 1 μm of the FOPO output can be measured by the autocorrelator (Femtocharome, FR-103XL). Figure 4.15 shows the autocorrelation waveforms of the amplified pump pulse and the idler of the FOPO output. It can be seen that the FWHM of the generated idler is compressed to 5.8 ps because of the pulse compression effect [20].

A high-efficiency all-fiber continuously tuned picosecond FOPO operating from 988 to 1046 nm and from 1085 to 1151 nm has been presented. The FWM signal gain band is located in the anomalous dispersion wavelength regime of the cavity. The parametric gain and the anomalous dispersion are provided by the 50-m PCF. However, the Hi-1060 SMF has large normal dispersion in the signal wavelength. A part of the anomalous dispersion intra-cavity is counteracted by the normal dispersion. It is hard to make full use of the cavity dispersion in this system. Then,

Fig. 4.15 Autocorrelation waveforms of the amplified pump pulse and the idler of the FOPO output. Reprinted from Ref. [20], copyright © OSA 2013, with permission from OSA

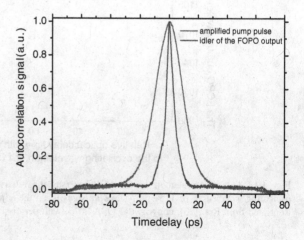

an optimized cavity with large normal dispersion in a wide wavelength range is investigated.

The optimized FOPO structure is shown in Fig. 4.16. A 10-m-long PCF is included in the cavity to provide enough parametric gain and relatively small anomalous dispersion. The ZDW of the PCF locates at 1062 nm, the nonlinear coefficient is 15 W^{-1} km^{-1}, and the fiber loss is measured to be 25 dB/km. To form a cavity with large normal dispersion, a section of Hi-1060 SMF is added to form a cavity with a length of 80 m. The optical isolators ISO1 and ISO2 are used to prevent unwanted feedback of optical radiations.

Figure 4.17 shows the optical parametric gain bands for the pump wavelength of 1061.8 nm with average pump power of 13 dBm injected into the PCF. Two optical

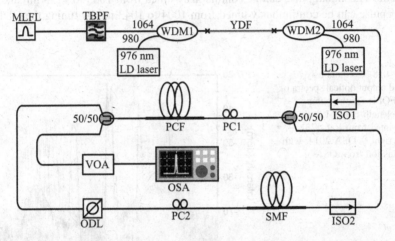

Fig. 4.16 Experimental setup of the optimized time-dispersion-tuned FOPO. Reprinted from Ref. [23], copyright © OSA 2014, with permission from OSA

Fig. 4.17 Optical parametric gain bands for the pump wavelength of 1061.8 nm. Reprinted from Ref. [23], copyright © OSA 2014, with permission from OSA

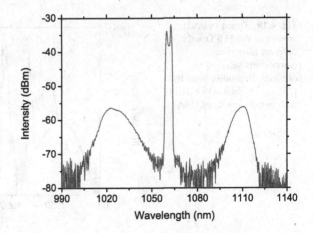

parametric gain bands can be observed locating on each side of the pump wave. And the bandwidth of the idler gain band is little broader than that of the signal gain band, due to the amplification of the amplified spontaneous emission (ASE) from the YDF amplifier. Then, the FOPO is built up by connecting the cavity. The spectral components in the optical parametric gain bands can be tuned to synchronize with the succedent pump pulse after a round-trip in the cavity by adjusting the cavity length. The round-trip time difference between different wavelength components is enlarged by the large normal cavity dispersion. The output wavelength of the FOPO can be continuously tuned by selecting different spectral components to oscillate. The measured output optical spectra of the FOPO with the pump wavelength fixed at 1061.8 nm are shown in Fig. 4.18, and the pump power is slightly adjusted to maintain the similar peak intensity and linewidth for all spectra. The signal pulse can be continuously tuned from 1083 to 1123 nm and the idler pulse can be continuously tuned from 1004 to 1043 nm by tuning the ODL in

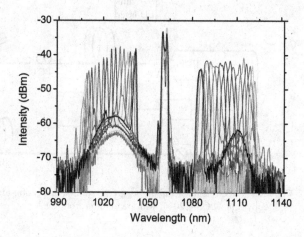

Fig. 4.18 Continuously tuned output optical spectra of the FOPO for the pump wavelength of 1061.8 nm. Reprinted from Ref. [23], copyright © OSA 2014, with permission from OSA

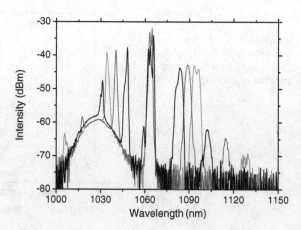

Fig. 4.19 Tuned optical spectra of the FOPO with different idler spectral components selected to oscillate. Reprinted from Ref. [23], copyright © OSA 2014, with permission from OSA

Fig. 4.20 Tuned optical spectra of the FOPO with different signal spectral components selected to oscillate. Reprinted from Ref. [23], copyright © OSA 2014, with permission from OSA

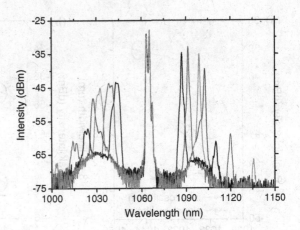

a range of about 24 mm, corresponding to 80 ps. The wavelength tuning range was roughly identical to the optical parametric gain bands [23].

In the FOPO configuration, either the signal or idler spectral components can be selected to oscillate. The tuned output optical spectra of the FOPO with the idler or signal components oscillated are shown in Figs. 4.19 and 4.20, respectively. When the idler spectral components are selected to oscillate, the idler intensity peaks are larger by about 4–6 dB than the corresponding signal intensity peaks, and the 3-dB linewidths of the idlers are narrower by about 2–3 nm than that of the corresponding signals. Inversely, when the signal spectral components are selected to oscillate, the signal intensity peaks are larger by about 8–11 dB than the corresponding idler intensity peaks, and the 3-dB linewidths of the signals are narrower by about 1–3 nm than that of the corresponding idlers, which are depicted in Fig. 4.21.

The output wavelength of the FOPO can be continuously tuned in a broad wavelength region by the combination of adjusting the pump wavelength and the time-dispersion-tuned technique as shown in Fig. 4.22. At first, the wavelength of 960 nm is oscillated, and the oscillating wavelength can be moved to longer wavelength region by increasing the cavity length. This is attributed to the normal cavity dispersion. The evolution of the output wavelength versus the relative optical delay time with the oscillating wavelength of 960 nm is shown in Fig. 4.23. Highly efficient and narrow linewidth outputs can be obtained in the wavelength regions from 960 to 1048 nm and from 1078 to 1180 nm by tuning the relative optical delay time in a range of 640 ps. At the output port of the FOPO, the phase-matched nonoscillating pulses are generated simultaneously. Except for the time dispersion tuning, the pump wavelength is also adjusted (1058, 1060, 1062, 1064, 1062, 1060, and 1058 nm), which are shown by the blue signs in Fig. 4.23.

In this section, normal and anomalous dispersion cavity structures are designed to synchronize the FWM parametric components with the pump pulses. The

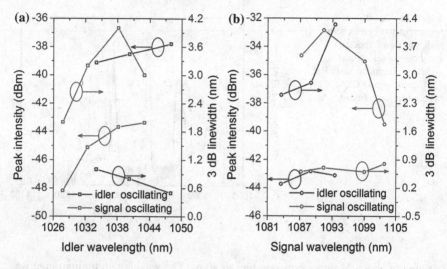

Fig. 4.21 Intensity peaks and 3-dB linewidths of idler (**a**) and signal (**b**) pulses for both of the signal or idler oscillating cases. Reprinted from Ref. [23], copyright © OSA 2014, with permission from OSA

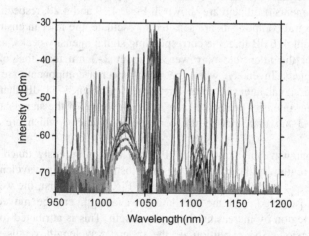

Fig. 4.22 Continuously tuned optical spectra of the FOPO built by the combination of adjusting the pump wavelength and the time-dispersion-tuned technique. Reprinted from Ref. [23], copyright © OSA 2014, with permission from OSA

output of the FOPO can be continuously tuned in a wide wavelength range by adjusting the pump wavelength in combination with the time-dispersion-tuned technique.

Fig. 4.23 Output wavelength of the FOPO versus the relative optical delay time with the oscillating wavelength of 960 nm. The blue signs mean the adjusting of the pump wavelength. Reprinted from Ref. [23], copyright © OSA 2014, with permission from OSA

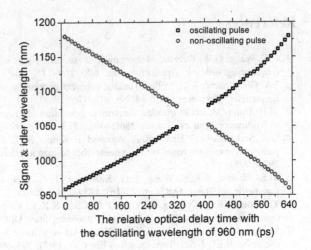

4.5 Conclusion

This chapter is focused to generate widely tunable light at new wavelength region by using the PCF-based OPO. The FWM parametric components generated in the PCF are coupled into the gain fiber repeatedly. Two kinds of wavelength tuning schemes are introduced. The first kind of schemes is focused to adjust the parameters in the phase-matching condition equation. The second kind of schemes is focused on designing cavity structures to make different wavelength component oscillated.

Firstly, a widely tunable FOPO is achieved by tuning the pump wavelength. According to the phase-matching condition, when the pump wavelength is tuned, both of the dispersion and nonlinear coefficient can be varied, and the central wavelength of the FWM parametric gain band can be tuned. A wavelength tuning range of 340 nm has been achieved.

Then, the time-dispersion-tuned technique is introduced to optimize the wavelength tunability of the FOPO. The parametric gain band can be broadened in time domain by a long cavity structure with large normal or anomalous dispersion. After a round-trip in the cavity, different parametric components can be selected to synchronize with the pump by slightly tuning the cavity length. It is easy to operate, and the wavelength tunability is largely optimized.

At last, the oscillated signal of the FOPO can continuously cover a wide wavelength band by adjusting the pump wavelength in combination with the time-dispersion-tuned technique. High-intensity and narrow linewidth output is achieved in a wide wavelength range.

References

1. C.C. Wang, G.W. Racette, Measurement of parametric gain accompanying optical difference frequency generation. Appl. Phys. Lett. **6**(8), 169–171 (1965)
2. J.A. Giordmaine, R.C. Miller, Tunable coherent parametric oscillation in LiNbO$_3$ at optical frequencies. Phys. Rev. Lett. **14**(24), 973–976 (1965)
3. M.H. Dunn, M. Ebrahimzadeh, Parametric generation of tunable light from continuous wave to femtosecond pulses. Science **286**(5444), 1513–1517 (1999)
4. D.K. Serkland, G.D. Bartolini, A. Agarwal, P. Kumar, W.L. Kath, Pulsed degenerate optical parametric oscillator based on a nonlinear-fiber Sagnac interferometer. Opt. Lett. **23**(10), 795–797 (1998)
5. M.E. Marhic, K.K.Y. Wong, L.G. Kazovsky, T.-E. Tsai, Continuous-wave fiber optical parametric oscillator. Opt. Lett. **27**(16), 1439–1441 (2002)
6. Y. Zhou, K.K.Y. Cheung, S. Yang, P.C. Chui, K.K.Y. Wong, Widely tunable picosecond optical parametric oscillator using highly nonlinear fiber. Opt. Lett. **34**(7), 989–991 (2009)
7. S. Yang, Y. Zhou, J. Li, K.K.Y. Wong, Actively mode-locked fiber optical parametric oscillator. IEEE J. Sel. Top. Quantum Electron. **15**(2), 393–398 (2009)
8. J.E. Sharping, M. Fiorentino, P. Kumar, R.S. Windeler, Optical parametric oscillator based on four-wave mixing in microstructure fiber. Opt. Lett. **27**(19), 1675–1677 (2002)
9. C.J.S. de Matos, J.R. Taylor, K.P. Hansen, Continuous-wave, totally fiber integrated optical parametric oscillator using holey fiber. Opt. Lett. **29**(9), 983–985 (2004)
10. Y. Deng, Q. Lin, F. Lu, G.P. Agrawal, W.H. Knox, Broadly tunable femtosecond parametric oscillator using a photonic crystal fiber. Opt. Lett. **30**(10), 1234–1236 (2005)
11. J.E. Sharping, J.R. Sanborn, M.A. Foster, D. Broaddus, A.L. Gaeta, Generation of sub-100-fs pulses from a microstructure-fiber-based optical parametric oscillator. Opt. Express **16**(22), 18050–18056 (2008)
12. M.E. Marhic, K.K.Y. Wong, L.G. Kazovsky, Wide-band tuning of the gain spectra of one-pump fiber optical parametric amplifiers. IEEE J. Sel. Top. Quant. Electron. **10**(5), 1133–1141 (2004)
13. J.D. Harvey, R. Leonhardt, S. Coen, G.K.L. Wong, J.C. Knight, W.J. Wadsworth, P.St. J. Russell, Scalar modulation instability in the normal dispersion regime by use of a photonic crystal fiber. Opt. Lett. **28**(22), 2225–2227 (2003)
14. K.C. Byron, M.A. Bedgood, A. Finey, C. McGauran, S. Savory, I. Watson, Shifts in zero dispersion wavelength due to pressure, temperature and strain in dispersion shifted singlemode fibres. Electron. Lett. **28**(18), 1712–1714 (1992)
15. G. Ghosh, H. Yajima, Pressure-dependent Sellmeier coefficients and material dispersions for silica fiber glass. J. Lightwave Technol. **16**(11), 2002 (1998)
16. L. Velazquez-Ibarra, A. Diez, E. Silvestre, M.V. Andres, M.A. Martinez, J.L. Lucio, Pump power dependence of four-wave mixing parametric wavelengths in normal dispersion photonic crystal fibers. IEEE Photon. Technol. Lett. **23**(14), 1010–1012 (2011)
17. J. Lasri, P. Devgan, R. Tang, J.E. Sharping, P. Kumar, A microstructure-fiber-based 10-GHz synchronized tunable optical parametric oscillator in the 1550-nm regime. IEEE Photon. Technol. Lett. **15**(8), 1058–1060 (2003)
18. R.H. Stolen, C. Lin, R.K. Jain, A time-dispersion-tuned fiber Raman oscillator. Appl. Phys. Lett. **30**(7), 340–342 (2008)
19. Y. Zhou, K.K.Y. Cheung, S. Yang, P.C. Chui, K.K.Y. Wong, A time-dispersion-tuned picosecond fiber-optical parametric oscillator. IEEE Photon. Technol. Lett. **21**(17), 1223–1225 (2009)
20. L. Zhang, S. Yang, P. Li, X. Wang, D. Gou, W. Chen, W. Luo, H. Chen, M. Chen, S. Xie, An all-fiber continuously time-dispersion-tuned picosecond optical parametric oscillator at 1 μm region. Opt. Express **21**(21), 25167–25173 (2013)
21. J.M. Dudley, G. Genty, S. Coen, Supercontinuum generation in photonic crystal fiber. Rev. Mod. Phys. **78**(4), 1135 (2006)

22. L. Zhang, S. Yang, X. Wang, D. Gou, X. Li, H. Chen, M. Chen, S. Xie, Widely tunable all-fiber optical parametric oscillator based on a photonic crystal fiber pumped by a picosecond ytterbium-doped fiber laser. Opt. Lett. **38**(22), 4534–4537 (2013)
23. L. Zhang, S. Yang, H. Chen, M. Chen, S. Xie, Broadly time-dispersion-tuned narrow linewidth all-fiber-integrated optical parametric oscillator, in *Optical Fiber Communication Conference, OFC*, 2014, W4E.2

Chapter 5
PCF-Based OPO with High Energy Conversion Efficiency

5.1 Introduction

The energy conversion efficiency is a critical factor for the practical application of the PCF-based OPO. As early as 1975, Stolen [1] demonstrated that the pump power can be effectively converted to the phase-matched signal and idler during the FWM process. An FOPA with a high gain over 70 dB has been reported [2], and many works have been performed to increase the energy conversion efficiency of the FOPO.

The FWM process requires that the fiber is uniform in the longitudinal axis. The fiber-based parametric gain can be largely decreased by the dispersion fluctuation along the length [3], so the highly efficient FOPO was firstly investigated in the telecommunication band. In the telecommunication band, the state-of-the-art HNLFs and DSFs can be used as a nonlinear medium. Wong et al. reported a FOPO with a high conversion efficiency over 40 %. The pump source was a tunable 4-ns pulse train formed from a CW laser modulated by a Lithium Niobate Mach-Zehnder intensity modulator. The walk-off between the pump and the signal was nearly vanished by the wide pulse width. The signal can be tuned from 1600 to 1860 nm, and the idler can be tuned from 1300 to 1500 nm by adjusting the pump wavelength near the ZDW [4]. Xu et al. [5] demonstrated a high-power FOPO in a short DSF. The oscillated signal was tuned to the wavelength located at the parametric gain peak. A maximal conversion efficiency of 93 % was obtained.

High conversion efficiency has been achieved in the FOPO by pumping in the telecommunication band. However, it is very hard to tune the output wavelength to other novel wavelength regions (1 μm and visible). The ZDW of the PCF can cover a quite wide wavelength range from the visible to the infrared band, so the PCF-based OPO can be potentially operated at any wavelength. The uniformity of the PCF is not as good as that of the DSF or HNLF, and the coupling loss (a coupling from free space to the PCF or a fusion between the PCF and the SMF) is larger than the fusion loss between the DSF and the SMF. The conversion efficiency

© Springer-Verlag Berlin Heidelberg 2016
L. Zhang, *Ultra-Broadly Tunable Light Sources Based on the Nonlinear Effects in Photonic Crystal Fibers*, Springer Theses,
DOI 10.1007/978-3-662-48360-2_5

of the PCF-based OPO is limited by these factors. In the last chapter, a PCF-based OPO can be operated in a wide wavelength range over 340 nm. The output wavelength can be continuously tuned by adjusting the pump wavelength combination with the time-dispersion-tuned technique. However, the energy conversion efficiency is very low (~ 1 %). In this chapter, we focused to increase the energy conversion efficiency of the PCF-based OPO through optimized cavity structures.

5.2   Doubly Resonant PCF-Based OPO

In the last chapter, we show that the intensity of the oscillated signal is larger than that of the newly generated idler, and the 3-dB bandwidth of the oscillated signal is narrower than that of the corresponding idler. It is clear that the amplified signal, which is synchronized with the pump, has a higher output power than the newly generated idler. In the FOPO investigated in this section, the signal and idler parametric components are controlled to synchronize with the pump simultaneously by a designed doubly resonant cavity structure. The performances of the signal and idler are expected to be enhanced by the doubly resonant cavity structure.

In the doubly resonant OPO structure, the signal and idler are fed back into the gain medium to oscillate simultaneously. In the singly resonant OPO, only one of the signals and the idler can be fed back into the gain medium, and the other is filtered out or not synchronized with the pump. The doubly resonant configuration has been applied in the crystal OPO to improve the output performance [6, 7]. In a single-longitudinal-mode FOPO operated in the telecommunication band, the threshold pump power is significantly reduced by using doubly resonant configuration [8]. Usually, the pulse-pumped FOPO can work in a wider wavelength region. However, the doubly resonant configuration is still not been used in the pulsed FOPO, to the best of our knowledge. It is desired to investigate the performance of the PCF-based OPO with doubly resonant configuration.

The experimental setup of the pulse-pumped FOPO with doubly resonant configuration is shown in Fig. 5.1. A homemade mode-locked ytterbium-doped picosecond fiber laser is used as the pump source. It can generate a pulse train with a FWHM of 35 ps at the repetition rate of 25.4 MHz. The output wavelength can be tuned from 1033 to 1070 nm. A bidirectional core-pumped ytterbium-doped fiber amplifier is set after a tunable band-pass filter with 1-nm 3-dB bandwidth to boost the pump power. After the amplifier, the pulse width is compressed to 21 ps. Subsequently, the pump pulse train is coupled into the cavity through the 50 % port of a fiber optical coupler. The polarization state of the pump wave can be adjusted by the polarization controller PC1. A PCF with a length of 10 m is included in the cavity to provide the parametric gain. The ZDW of the PCF locates at 1062 nm, and the nonlinear coefficient is calculated to be about 15 $W^{-1} km^{-1}$. The fusion loss between the PCF and the Hi-1060 SMF is minimized by inserting a piece of ultra-high numerical aperture fiber between them. After the PCF, a fiber optical coupler provides 50 % output and 50 % feedback. The light in the feedback branch is split into

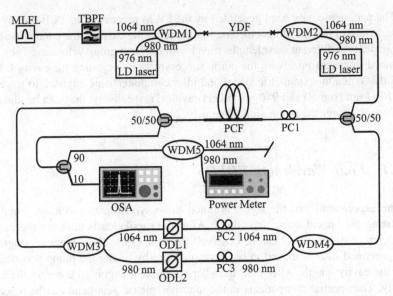

Fig. 5.1 Experimental setup of the doubly resonant FOPO. *MLFL* mode-locked ytterbium-doped fiber laser; *TBPF* tunable band-pass filter; *WDM* wavelength division multiplex; *YDF* ytterbium-doped fiber; *PC* polarization controller; *PCF* photonic crystal fiber; *ODL* optical delay line; *OSA* optical spectrum analyzer

two light paths by the 980-nm port and the 1060-nm port of the WDM3. Then, the two optical paths are combined by the WDM4. Two oscillating cavities with a similar length of about 24 m are formed. In each cavity, an optical delay line is used to precisely control the cavity length, and a polarization controller is used to adjust the polarization state of the light in the cavity. When the wavelength of the idler is less than 1 μm, it is guided through the 980-nm path, and the pump and signal components are guided through the 1060-nm path. The losses of the idler and signal cavities are measured to be 9 and 8.5 dB, respectively. At the output port, an optical spectrum analyzer (OSA) is used to record the spectrum. The idler components below 1 μm can be filter out by the WDM5, and the power can be measured [9].

The oscillation of the FOPO is based on the optical parametric process. The parametric gain bands (signal and idler) can be generated by pumping near the ZDW of the fiber. The gain bandwidth is governed by Eq. (2.8). The central wavelengths of the signal and idler gain bands are determined by the phase-matching condition, and the locations can be tuned by adjusting the pump wavelength. The parametric gain components can be selected to feed back into the gain fiber, the selected components can be amplified by the synchronized pump, and the corresponding idler components can be generated simultaneously. When the parametric gain is increased to be larger than the cavity loss, a dynamic stability can be formed, and the FOPO begins to oscillate.

The parametric gain band generated by the FWM process in the PCF usually has a bandwidth of dozens of nanometers. In the oscillating cavities, the parametric gain components at different wavelengths travel with different group velocities. They can be tuned to synchronize with the pump successively by adjusting the cavity length slightly. When the parametric signal and idler components are adjusted to travel in the 1060-nm (signal) and 980-nm (idler) cavities, respectively, they can be tuned to synchronize with the pump simultaneously.

5.2.1   Four-Wavelength FOPO

In the experiment, first the signal is tuned to be synchronized with the pump by adjusting the optical delay line ODL1. After the oscillation is built up, the output spectrum of the oscillator can be observed from the OSA. Then, the wavelength of the generated idler is moved to be less than 1 μm by tuning the pump wavelength and the cavity length. After that, the idler component begins to travel in the idler cavity. The spectral components in the idler parametric gain band can be tuned to oscillate by tuning the ODL2 in the idler cavity. When a wavelength in the signal cavity and a wavelength in the idler cavity, which are phase mismatched with each other, are tuned to synchronize with the pump simultaneously, a four-wavelength output can be obtained [9].

One of the output optical spectra of the four-wavelength FOPO is shown in Fig. 5.2a. The pump locates at 1059.5 nm, the wavelengths of the two signals are 1153.5 and 1159.5 nm, and the wavelengths of the two idler are 976.3 and 981.3 nm. The lobe located from 1015 to 1045 nm is from the amplified spontaneous emission generated in the ytterbium-doped fiber amplifier. The gaps between the two signals and the two idlers can be tuned by adjusting the ODL1 or ODL2. Figure 5.2b shows another optical spectrum of the four-wavelength FOPO with the pump wavelength fixed at 1059.5 nm. The two signal wavelengths are tuned to 1136 and 1153 nm, and the two idler wavelengths are tuned to 981 and 995 nm. It can be inferred that the FOPO can also be operated in the 8-wavelength, 16-wavelength, and 32-wavelength style by utilizing the multiple cavity structure.

5.2.2   Doubly Resonant FOPO with High Conversion Efficiency

In a doubly resonant FOPO, the spectral components in the parametric signal and idler gain bands can be tuned to oscillate simultaneously by adjusting the signal and idler cavity lengths, respectively. In the last section, a four-wavelength FOPO is formed, and the frequency gap between the two signals and the two idlers can be tuned by adjusting the ODL1 or ODL2. When the two signals and the two idlers are

Fig. 5.2 The optical spectra of the four-wavelength FOPO with the pump fixed at 1059.5 nm; the two signals and the two idlers can be tuned

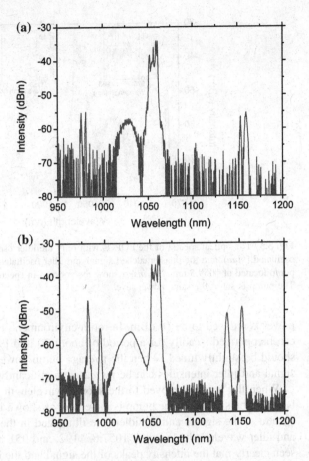

adjusted to be overlapped, the signal- and the phase-matched idlers are selected to oscillate simultaneously. Figure 5.3 shows the optical spectra of the FOPOs with either the signal or the idler oscillated. The pump is set at the same power level. The green line denotes only the idler oscillated, the blue line denotes only the signal oscillated, and the red line denotes the phase-matched signal and idler oscillated simultaneously. The zoomed-in spectra of the signals and the idlers are illustrated in the insets of Fig. 5.3. It can be seen that once the signal- and the phase-matched idler are tuned to oscillate simultaneously, the output signal and idler intensities are increased immediately. Compared to the optical spectrum of the doubly resonant oscillator with that of the singly resonant case, the signal intensity peak is increased by 3.4–5 dB, the idler intensity peak is increased by 2.9–4.4 dB, the 3-dB bandwidth of the signal is compressed by 0.96–1.73 nm, and the 3-dB bandwidth of the idler is compressed by 0.09–0.42 nm. In the FOPO with only the idler oscillated, the threshold average pump power coupled into the PCF is measured to be 12 dBm. In the FOPO with only the signal oscillated, the threshold average pump power is measured to be 11.5 dBm. In the doubly resonant case, the threshold average pump

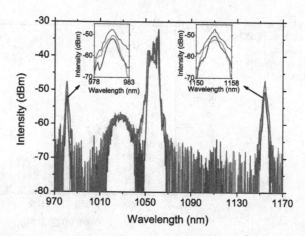

Fig. 5.3 The optical spectra of the FOPOs with only the idler oscillated (*green*), only the signal oscillated (*blue*), and the phase-matched signal and idler oscillated simultaneously (*red*) with the pump located at 1059.5 nm. The *insets* show the zoomed-in spectra of the signals and the idlers. The pump is set at the same power level

power is reduced to be 10 dBm. In room environment, the doubly resonant FOPO can be operated steadily for a period of about 20 min [9]. After 20 min, the ODL should be slightly tuned. When the average pump power is increased slightly, the signal and idler intensities can be increased significantly, as shown in Fig. 5.4.

When the pump is moved to the shorter wavelength region, the frequency gap between the signal and the pump is increased, as shown in Fig. 5.5. The zoomed-in spectra of the signals and the idlers are illustrated in the insets. The pump, signal, and idler wavelengths locate at 1057.6, 1192, and 951 nm, respectively. It can be seen clearly that the intensity peaks of the signal and the idler in the FOPO with the signal synchronized with the pump are larger than that of the FOPO with the idler synchronized with the pump. It is because the loss of the signal cavity is smaller

Fig. 5.4 The output spectrum of the doubly resonant FOPO by increasing the pump power over the threshold value

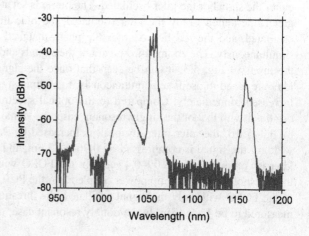

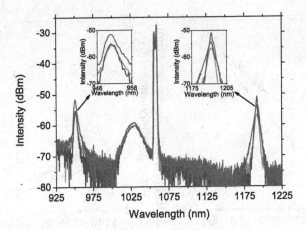

Fig. 5.5 The optical spectra of the FOPOs with only the idler oscillated (*green*), only the signal oscillated (*blue*), and the phase-matched signal/idler oscillated simultaneously (*red*) with the pump located at 1057.6 nm. The *insets* show the zoomed-in spectra of the signals and the idlers. The pump is set at the same power level

than that of the idler cavity. Compared to the doubly resonant FOPO with the singly resonant FOPOs, the intensity peaks of the signal and idler are increased and the 3-dB bandwidths of the signal and idler are compressed. Figure 5.6 shows the increment of the intensity peak and the decrement of the 3-dB bandwidth versus the wavelength. In the case of comparing with the singly resonant FOPO with only the idler oscillated, for the idler ranging from 992 to 951 nm, the intensity peak increment increases from 2.64 to 4.18 dB, and the 3-dB bandwidth decrement increases from 0.05 to 0.5 nm. For the signal from 1140 to 1192 nm, the intensity peak increment increases from 4.96 to 5.77 dB, and the 3-dB bandwidth decrement increases from 1.5 to 2.98 nm. In the case of comparing with the FOPO with only the signal oscillated, for the idler from 992 to 951 nm, the intensity peak increment decreases from 4.83 to 3.66 dB, and the decrement of the 3-dB bandwidth decreases from 0.6 to 0.16 nm. For the signal from 1140 to 1192 nm, the increment of the intensity peak decreases from 3.7 to 2.53 dB, and the decrement of the 3-dB bandwidth decreases from 1.12 to 0.51 nm [9].

For the idler component below 1 μm filtered by the WDM5 at the output port of the FOPO, the power can be measured. Figure 5.7 shows the average idler power of the singly and doubly resonant FOPOs versus the wavelength. The average pump power coupled into the PCF is set at 15.5 dBm. In the singly resonant FOPO with only the idler oscillated, for the idler from 951 to 992 nm, the output average power increases from −6.5 to −3.95 dBm, and the energy conversion efficiency increases from 0.63 to 1.13 %. In the singly resonant FOPO with only the signal oscillated, for the idler from 951 to 992 nm, the output average power increases from −6 to −4.2 dBm, and the energy conversion efficiency increases from 0.71 to 1.07 %. In the doubly resonant FOPO with the phase-matched signal/idler oscillated simultaneously, for the idler from 951 to 992 nm, the output average power increases from

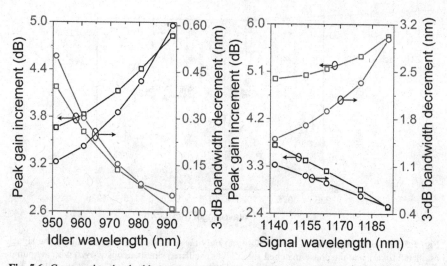

Fig. 5.6 Compared to the doubly resonant oscillator with the singly resonant oscillators with only the idler or only the signal synchronized with the pump, the signal and idler intensity peaks increment and 3-dB bandwidth decrement as a function to the wavelength

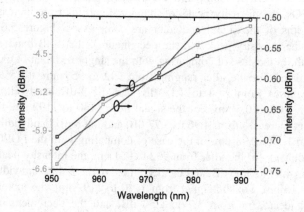

Fig. 5.7 The output average idler powers of the FOPOs with only the idler oscillated (*green*), only the signal oscillated (*blue*), and the phase-matched signal/idler oscillated simultaneously (*red*) as a function of the wavelength

−0.72 to −0.51 dBm, and the energy conversion efficiency increases from 2.39 to 2.51 %. It can be seen that the energy conversion efficiency can be improved by 2.2–3.8 times by the doubly resonant cavity structure [9].

In this section, a PCF-based OPO with doubly resonant configuration has been demonstrated. When a wavelength in the signal cavity and a wavelength in the idler cavity (phase mismatched with the oscillated signal wavelength) are tuned to oscillate simultaneously, a four-wavelength FOPO can be achieved. When the phase-matched signal and phase-matched idler are tuned to oscillate simultaneously, the doubly resonant FOPO begins to work. To compare the doubly resonant FOPO with the singly resonant FOPOs, the intensity peaks of the output signal and idler are increased by 2.53 to 5.77 dB, the 3-dB bandwidths of the signal and the idler are compressed by 0.05 to 2.98 nm, and the energy conversion efficiencies are improved by 2.2 to 3.8 times.

5.3 Highly Efficient PCF-Based OPO with All-Fiber Cavity Structure

In the doubly resonant FOPO reported in the last section, both the signal and the idler intensities are enhanced, and the energy conversion efficiencies are improved by 2.2 to 3.8 times with respect to the singly resonant FOPO. However, the conversion efficiency is still in the level of about 3 %. To compare with the highly efficient FOPOs operated at the telecommunication band, the conversion efficiency is still very low. After a careful analysis about the oscillating system, we think that the cavity loss has critical influence on the conversion efficiency.

In order to explore the possibility to build a high-efficiency PCF-based OPO, an all-fiber cavity structure is designed to minimize the cavity loss. The experimental setup is shown in Fig. 5.8. In the previous PCF-based OPO systems, a free-space optical time delay line worked at the telecommunication band is inserted into the cavity to adjust the cavity length. However, it introduces about 4 dB loss in the 1 μm region. In this system, a specific length of Hi-1060 SMF instead of the free-space optical delay line is used to adjust the cavity length.

The cavity loss is minimized as far as possible by the elimination of any free-space devices in the oscillating cavity. The threshold peak pump power coupled into the PCF is decreased to about 10 W. When the pump power exceeds the

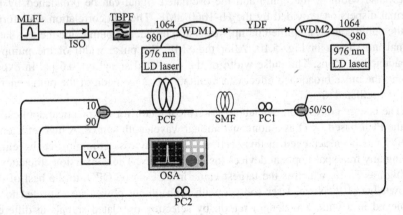

Fig. 5.8 The experimental setup of the PCF-based OPO with all-fiber cavity structure

Fig. 5.9 The optical spectra of the high conversion efficiency PCF-based OPO for the pump wavelengths of 1062 and 1060 nm

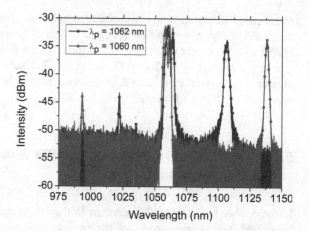

threshold power, a large portion of the pump power is transferred to the oscillated signal wavelength. The output spectra of the FOPO are illustrated in Fig. 5.9. After the pump wavelength tuned from 1060 to 1062 nm, the oscillating signal wavelength is moved from 1138 to 1107 nm. The intensity peak of the oscillating signal is slightly smaller than the intensity peak of the residual pump, and the energies near the central wavelength of the pump are severely depleted, which means that the FOPO has high energy conversion efficiency. At the output port of the FOPO, the pump, signal, and idler can be split by a diffraction grating. For the pump wavelength of 1060 nm, the energy conversion efficiency from the pump to the signal is about 30 %. For the pump wavelength of 1062 nm, 33 % of the pump power is converted into the oscillated signal, and 3 % of the pump power is converted into the new generated idler. A total conversion efficiency (from the pump to the signal and the idler) of 36 % can be obtained. The external energy conversion efficiency (output signal/idler power/input pump power) is calculated to be about 16 % [10].

In the configuration of the FOPO, a spool of Hi-1060 SMF is set before the PCF. The pulse width of the pump and the oscillated signal can be broadened by the normal dispersion provided by the Hi-1060 SMF. The autocorrelation waveforms of the pump source, amplified pump, dispersion broadened pump, and the oscillated signal are shown in Fig. 5.10. After the SMF, the pulse width of the pump is broadened to 25 ps. The pulse width of the oscillated signal is 10.6 ps. In experiment, the pulse broadened effect can facilitate the conversion of the pump energy to the oscillated signal.

The energy conversion efficiency is a critical factor for the practical application of the PCF-based OPO as a nonconventional wavelength source. A high-efficiency FOPO has been achieved in this section. The cavity loss is minimized by eliminating any free-space optical devices inside the cavity. The conversion efficiency is as high as 36 %, which is the largest value in PCF-based OPO, to the best of our knowledge [10]. Novel laser sources with high output power are expected to be achieved in a wide wavelength region by selecting oscillated signals at different wavelengths.

Fig. 5.10 The autocorrelation waveforms of the original pump, the amplified pump, the dispersion broadened pump, and the oscillated signal

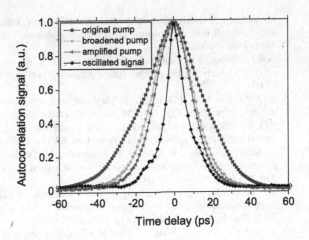

5.4 Conclusion

This chapter is focused on increasing the energy conversion efficiency of the PCF-based OPO. Two kinds of cavity structures are proposed.

Based on the experience that the oscillated signal has a higher intensity than the new generated idler in the FOPO, a doubly resonant configuration is used in a PCF-based OPO. When a wavelength in the signal cavity and a wavelength in the idler cavity (phase mismatched with the oscillated signal wavelength) oscillate simultaneously, a four-wavelength output can be achieved. When the phase-matched signal and idler oscillate simultaneously, the doubly resonant FOPO begins to work. Compared to the doubly resonant FOPO with the singly resonant FOPOs, the intensity peaks of the output signal and idler are increased, the 3-dB bandwidths of the signal and the idler are compressed, and the energy conversion efficiencies can be improved by 2.2 to 3.8 times.

Although the conversion efficiency can be improved by the doubly resonant configuration, it is still much smaller than that of the highly efficient FOPO operated at the telecommunication band. Then, an all-fiber cavity structure is used to enhance the energy conversion efficiency. The cavity loss is minimized by eliminating any free-space optical devices inside the cavity. The conversion efficiency is as high as 36 %, which is the largest value in PCF-based OPO, to the best of our knowledge.

References

1. R.H. Stolen, Phase-matched-stimulated four-photon mixing in silica-fiber waveguides. IEEE J. Quantum Electron. **11**(3), 100–103 (1975)
2. T. Torounidis, P.A. Andrekson, B.E. Olsson, Fiber-optical parametric amplifier with 70-dB gain. IEEE Photon. Technol. Lett. **18**(10), 1194–1196 (2006)

3. J.S.Y. Chen, S.G. Murdoch, R. Leonhardt, J.D. Harvey, Effect of dispersion fluctuations on widely tunable optical parametric amplification in photonic crystal fibers. Opt. Express **14**(20), 9491–9501 (2006)

4. G.K.L. Wong, S.G. Murdoch, R. Leonhardt, J.D. Harvey, V. Marie, High-conversion-efficiency widely tunable all-fiber optical parametric oscillator. Opt. Express **15**(6), 2947–2952 (2007)

5. Y.Q. Xu, K.F. Mak, S.G. Murdoch, Multiwatt level output powers from a tunable fiber optical parametric oscillator. Opt. Lett. **36**(11), 1966–1968 (2011)

6. F.G. Colville, M.J. Padgett, M.H. Dunn, Continuous-wave, dual-cavity, doubly resonant, optical parametric oscillator. Appl. Phys. Lett. **64**(12), 1490–1492 (1994)

7. B. Scherrer, I. Ribet, A. Godard, E. Rosencher, M. Lefebvre, Dual-cavity doubly resonant optical parametric oscillators: demonstration of pulsed single-mode operation. J. Opt. Soc. Am. B **17**(10), 1716–1729 (2000)

8. S. Yang, X. Xu, Y. Zhou, K.K.Y. Cheung, K.K.Y. Wong, Continuous-wave single-longitudinal-mode fiber-optical parametric oscillator with reduced pump threshold. IEEE Photon. Technol. Lett. **21**(24), 1870–1872 (2009)

9. L. Zhang, S. Yang, X. Wang, D. Gou, H. Chen, M. Chen, S. Xie, Picosecond photonic crystal fiber-based doubly resonant optical parametric oscillator. IEEE Photon. Technol. Lett. **26**(7), 682–685 (2014)

10. L. Zhang, S. Yang, X. Wang, D. Gou, H. Chen, M. Chen, S. Xie, High-efficiency all-fiber optical parametric oscillator based on photonic crystal fibers pumped by ytterbium-doped fiber laser. Electron. Lett. **50**(8), 624–626 (2014)

Chapter 6
Conclusion

This thesis is focused on the investigation of the widely tunable laser generation at novel wavelength bands through the nonlinear optical process in the photonic crystal fibers (PCFs). The PCF-based nonlinear effects of four-wave mixing (FWM), dispersive wave (DW) generation, and cross-phase modulation (XPM) are analyzed and used. The required PCFs are designed and fabricated. Based on the FWM optical parametric gain generated in the PCFs, the very promising optical devices of PCF-based optical parametric amplifier (OPA) and optical parametric oscillator (OPO) are investigated. The main contents are listed below:

1. The influences of the dispersion, nonlinear coefficient, and pump wavelength on the FWM parametric gain spectrum generated in the PCF are theoretically investigated. Based on the parameters of the PCF, the shape of parametric gain spectrum at different pump wavelengths can be predicted accurately. The structure of the PCF with two zero-dispersion wavelengths (TZDWs) is designed by using the multipole method. The signal at the telecommunication band can be converted to the blue/green visible wavelength band. The expected large span wavelength conversion is achieved in the fabricated PCF with TZDW by using a Ti: sapphire laser as the pump. Furthermore, visible and mid-infrared DWs are generated in the PCF by pumping in the anomalous dispersion wavelength regime. The visible DW can be generated from 498 nm to 425 nm, and the mid-infrared DW can be tuned from 1986 to 2279 nm. In the PCF with TZDW, the spectrum can be extended to the ultraviolet region from 200 to 400 nm by the XPM between the anti-Stokes signal and the Raman soliton.

2. In order to use the mode-locked ytterbium-doped fiber laser as the pump to form a widely tunable light source, a dispersion-flattened PCF with the ZDW located near 1064 nm is designed and fabricated. Widely tunable optical parametric gain sidebands are generated in the fabricated PCF. The picosecond pulse laser is proposed as the pump source to build the PCF-based optical parametric amplifier, and the method to calculate the parametric gain for the pulse-pumped FOPA is developed. High gain can be obtained in a broadband by tuning the

© Springer-Verlag Berlin Heidelberg 2016

L. Zhang, *Ultra-Broadly Tunable Light Sources Based on the Nonlinear Effects in Photonic Crystal Fibers*, Springer Theses,

DOI 10.1007/978-3-662-48360-2_6

pump wavelength of the PCF-based OPA. The CW signal can be amplified to picosecond pulse by the pulsed pump, and the picosecond idler can also be generated. The pulse widths of the signal and idler are narrower than that of the pump. The picosecond pulse can be generated in a wide wavelength band by tuning the wavelengths of the pump and the signal.

3. The wavelengths of the signal and idler pulses generated in the FOPA are limited by the accessible signal source. In order to get rid of the limitation, a ring cavity, which can feed the FWM parametric components repeatedly into the gain medium, is used to form the PCF-based OPO. The PCF-based OPO can be operated in a wide wavelength range over 340 nm by tuning the pump wavelength. When the pump wavelength is tuned about only one nanometer in the normal dispersion regime, the output wavelength of the FOPO can be detuned by tens of nanometers. It is very difficult to tune the pump wavelength rigorously so as to continuously tune the output wavelength of the FOPO. In order to optimize the wavelength tunability of the FOPO, the time-dispersion-tuned technique is introduced. High intensity and narrow linewidth output can be achieved in a wide wavelength band by adjusting the pump wavelength in combination with the time-dispersion-tuned technique.

4. The energy conversion efficiency is a critical factor for the practical application of the PCF-based OPO. Two kinds of cavity structures are designed to increase the conversion efficiency of the PCF-based OPO. In the FOPO, the oscillated signal has a higher intensity than the newly generated idler, so a doubly resonant configuration is used in the PCF-based OPO. When a wavelength in the signal cavity and a wavelength in the idler cavity (phase mismatched with the oscillated signal wavelength) are tuned to oscillate simultaneously, a four-wavelength output can be achieved. After the phase-matched signal/idler oscillated simultaneously, the doubly resonant FOPO begins to work. Comparing the doubly resonant FOPO with the singly resonant FOPOs, the intensity peaks of the output signal and idler are increased, the 3-dB bandwidths of the signal and idler are compressed, and the energy conversion efficiencies can be improved by 2.2–3.8 times. In order to further increase the conversion efficiency, the all-fiber cavity is used in the PCF-based OPO. A high conversion efficiency of 36 % is obtained, which is the largest value in PCF-based OPO, to the best of our knowledge.

Printed in the United States
By Bookmasters